Chasing Water

CHASING WATER

A Guide for Moving from Scarcity to Sustainability

Brian Richter

Washington | Covelo | London

Library of Congress Cataloging-in-Publication Data
Richter, Brian D.
Chasing water : a guide for moving from scarcity to sustainability / Brian Richter.
pages cm
Includes bibliographical references and index.
ISBN-13: 978-1-61091-536-6 (hardback)
ISBN-10: 1-61091-536-4 (cloth)
ISBN-13: 978-1-61091-538-0 (paper)
1. Water supply—Government policy. 2. Water supply--International cooperation. 3. Water resources development—Government policy. 4. Water resources development--International cooperation. I. Title.
HD1691.R53 2014
333.91—dc23

2013043021

Printed on recycled, acid-free paper

Manufactured in the United States of America
10 9 8 7 6 5 4 3 2 1

Island Press would like to thank the New Mexico Water Initiative, a project of the Hanuman Foundation, for generously supporting the production of this book.

Keywords
Island Press, water scarcity, water shortage, water stress, water conservation, water governance, water entitlements, water rights, water allocation, water budget, water economics, water markets, water sustainability, water policy, water privatization, water planning, droughts, desalination, water importation, water transfers, water withdrawals, consumptive use, irrigation efficiency, water reuse, water recycling, reservoirs, watershed, aquifer, groundwater, water cycle, river conservation, freshwater ecosystems, environmental flow, High Plains Aquifer, Murray-Darling Basin, Colorado River

For Walt and Ann,

who showed me the world and encouraged me

to make a difference in it

For Martha,

whose love keeps me on course

And for Henry,

who will inherit the world we create

I was born in a drought year. That summer my mother waited in the house, enclosed in the sun and the dry ceaseless wind, for the men to come back in the evenings, bringing water from a distant spring. Veins of leaves ran dry, roots shrank. And all my life I have dreaded the return of that year, sure that it still is somewhere, like a dead enemy's soul. Fear of dust in my mouth is always with me, and I am the faithful husband of the rain. I love the water of wells and springs and the taste of roofs in the water of cisterns. I am a dry man whose thirst is praise of clouds, and whose mind is something of a cup. My sweetness is to wake in the night after days of dry heat, hearing the rain.

Wendell Berry, "Water" poem in
Farming: A Handbook

About Island Press

Since 1984, the nonprofit organization Island Press has been stimulating, shaping, and communicating ideas that are essential for solving environmental problems worldwide. With more than 800 titles in print and some 40 new releases each year, we are the nation's leading publisher on environmental issues. We identify innovative thinkers and emerging trends in the environmental field. We work with world-renowned experts and authors to develop cross-disciplinary solutions to environmental challenges.

Island Press designs and executes educational campaigns in conjunction with our authors to communicate their critical messages in print, in person, and online using the latest technologies, innovative programs, and the media. Our goal is to reach targeted audiences—scientists, policymakers, environmental advocates, urban planners, the media, and concerned citizens—with information that can be used to create the framework for long-term ecological health and human well-being.

Island Press gratefully acknowledges major support of our work by The Agua Fund, The Andrew W. Mellon Foundation, Betsy & Jesse Fink Foundation, The Bobolink Foundation, The Curtis and Edith Munson Foundation, Forrest C. and Frances H. Lattner Foundation, G.O. Forward Fund of the Saint Paul Foundation, Gordon and Betty Moore Foundation, The Kresge Foundation, The Margaret A. Cargill Foundation, New Mexico Water Initiative, a project of Hanuman Foundation, The Overbrook Foundation, The S.D. Bechtel, Jr. Foundation, The Summit Charitable Foundation, Inc., V. Kann Rasmussen Foundation, The Wallace Alexander Gerbode Foundation, and other generous supporters.

The opinions expressed in this book are those of the author(s) and do not necessarily reflect the views of our supporters.

Advance praise for *Chasing Water*

"With clarity and vision, Brian Richter has written the perfect primer on water scarcity—what it is, where it exists, and what we can do about it. It will be of immense value to students, professors, policy makers, and anyone interested in creating a more secure water future."

—Sandra Postel, Freshwater Fellow, National Geographic Society

"How the global community manages and cares for this most precious resource will be one of the most urgent questions in the twenty-first century. This powerful and absorbing book offers principles for citizen-empowered water governance and stewardship that integrate broad societal values. This is an indispensable book for policy makers and citizens of the world."

—Gavin Power, Executive Director,
United Nations CEO Water Mandate

"Richter distills the complex global dynamic of expanding water scarcity into an eloquent personal narrative that explains the challenges and offers practical tools to tackle them. With straightforward solutions that apply equally to the southwestern United States and sub-Saharan Africa, Richter empowers his readers and inspires action. Readers from many backgrounds will benefit from this book, and I count myself among them."

—Michael McClain, Professor , UNESCO-IHE Institute
of Water Education, Delft, Netherlands

"*Chasing Water* is a cogent, hopeful vision of achieving water sustainability for humans and aquatic ecosystems alike, from a global expert on freshwater policy and conservation science. Richter's sustainability principles and case studies illustrating democratic water governance inspire us to seek collective solutions for seemingly intractable water management challenges."

—LeRoy Poff, Professor of Biology, Colorado State University

"*Chasing Water* is a must-read for all river enthusiasts and seekers of practical solutions. Immensely readable and intriguing, this book will awaken your senses, provoke you to think deeply about the choices we make, and give you tools to carve a more sustainable future. Don't miss it!"

—Nicole Silk, President, River Network

Contents

Chasing Water

Places Discussed in this Book

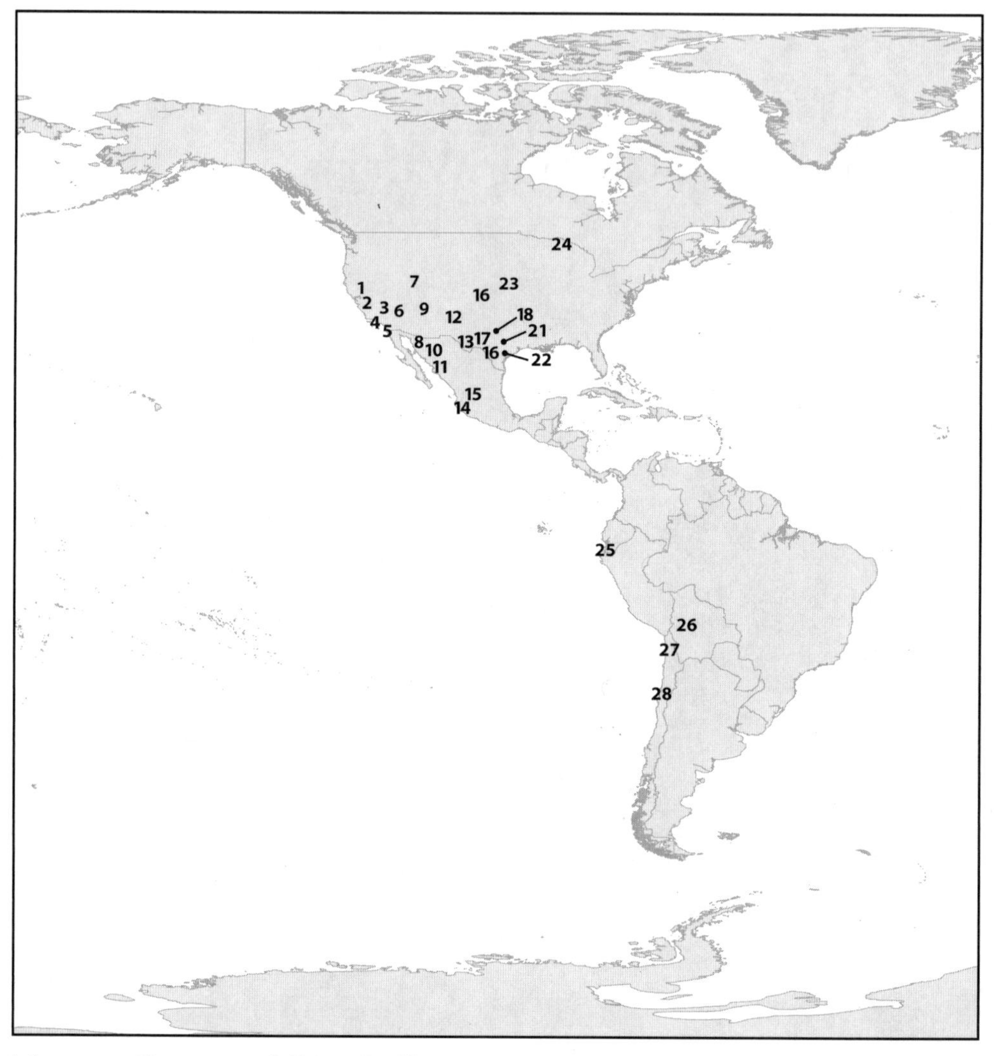

1 Sacramento River
2 San Joaquin River and Central Valley Aquifer
3 Owens Lake
4 Los Angeles
5 San Diego
6 Las Vegas
7 Great Salt Lake
8 Conception River
9 Colorado River (Pacific Ocean)
10 Western Mexico Aquifer
11 Fuerte River
12 Acequias of New Mexico
13 Rio Grande
14 Armeria River
15 Santiago River
16 High Plains Aquifer
17 Colorado River (Texas)
18 Brazos River
19 Guadalupe-San Antonio River
20 Edwards Aquifer
21 Austin
22 San Antonio
23 Missouri and Mississippi Rivers
24 Lake Superior
25 Chira River
26 Cochabamba
27 Loa River
28 Huasco River

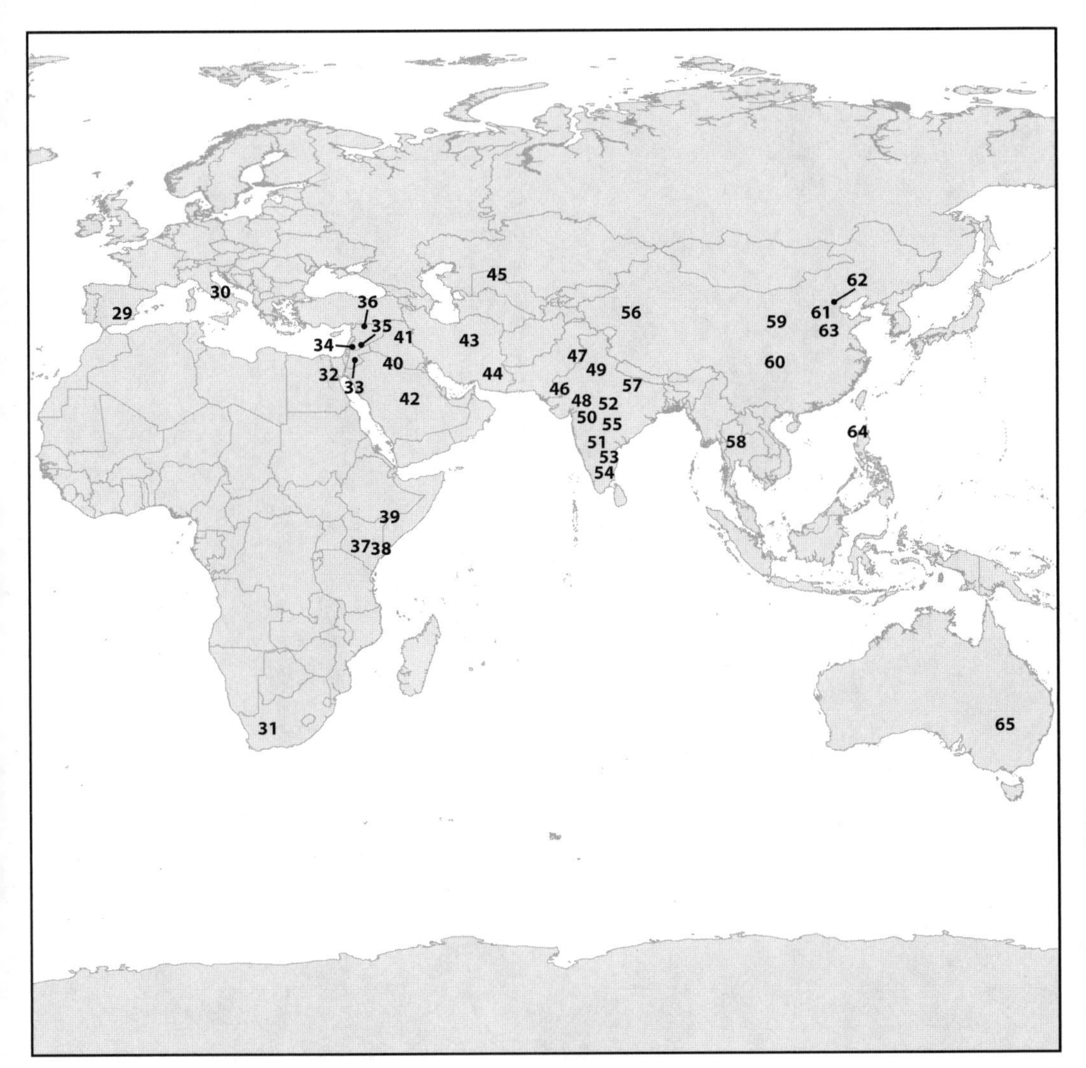

29 Guadiana River
30 Tiber River
31 Doring River
32 Nile Delta Aquifer
33 Dead Sea
34 Jordan River
35 Damascus
36 Aleppo
37 Lake Naivasha
38 Tana River
39 Shebelle River
40 North Arabian Aquifer
41 Tigris-Euphrates
42 Saudi Arabia
43 Persian Aquifer
44 South Caspian Aquifer
45 Aral Sea
46 Lower Indus Aquifer
47 Indus River
48 Mahi River
49 Upper Ganges Aquifer
50 Tapti River
51 Krishna River
52 Narmada River
53 Penner River
54 Cauvery River
55 Godavari River
56 Tarim River
57 Ganges River
58 Chao Phraya
59 Yellow River
60 Yangtze River
61 Yongding River
62 Beijing
63 North China Plain Aquifer
64 Bacarra-Vintar River
65 Murray-Darling Basin

CHAPTER ONE

Running Out of Water

In March 1934, Benjamin Baker Moeur, then governor of the state of Arizona in the United States, became extremely agitated upon hearing that the neighboring state of California was preparing to build a dam on the Colorado River to deliver more water to growing cities in Southern California. The river, which in its lower reaches forms the border between California and Arizona, had recently shriveled to a fifth of its normal water flow after 5 years of parching drought across the western part of the country.

Moeur had not been advised of any plans to build a dam on the lower Colorado, and he worried that California was going to take more than its fair share of a river that was already showing signs of strain.

Upon hearing the news of the dam's construction, the governor immediately sent six members of the Arizona National Guard to investigate. They traveled upriver from Yuma aboard a ramshackle ferryboat named the *Nellie Jo*, newly christened as the "Arizona Navy" for its reconnaissance mission. In an escapade reminiscent of the Three Stooges comedies that began appearing in movie theaters that same year, the *Nellie Jo* ran aground on a sandbar just below the dam construction site, and the guardsmen had to be rescued by construction workers.

The guardsmen continued to monitor the dam work for 7 months, sending daily dispatches to the governor by radio. When the guardsmen reported that construction activities had finally touched upon Arizona's shore, Moeur became incensed. He invoked martial law, issued a proclamation to "Repel an Invasion," and sent out a 100-man militia unit in eighteen trucks, some with mounted machine guns, to halt construction.[1]

Another seventeen truckloads of troops were prepared to head upriver in a ferryboat flotilla when U.S. Secretary of the Interior Harold Ickes stepped in, asking Moeur to back down, and pacifying him with assurances that all dam construction would be stopped until the dispute could be settled.

But Ickes was livid over Moeur's aggressive actions, and he retaliated by suing Arizona in the U.S. Supreme Court to stop the state from interfering with the construction of Parker Dam. After all, he reasoned, California's right to divert water from the Colorado had been explicitly authorized by Congress in 1922 through the Colorado River Compact, also known as the "Law of the River," which allocated portions of the river to each of the seven states through which the river flows.

Much to Ickes's surprise, the Supreme Court ruled in Arizona's favor. The court noted that Parker Dam, to be built by a federal water agency, had never been formally authorized by the U.S. Congress, and Arizona had not yet signed the Colorado River Compact, because it disagreed with the small amount of water it was being given through the compact.

Eventually, Moeur dropped his opposition to Parker Dam in exchange for a commitment from Ickes to provide federal funding for substantially expanding irrigation projects in Arizona with water from the Colorado. Arizona subsequently signed the Colorado River Compact in 1944.

The drought of the 1930s came and went, but water development projects sanctioned in that decade fixed the fate of the Colorado River for the century that would follow (fig. 1-1). Hoover Dam, when completed in 1936, became the largest dam in the world at the time. Parker Dam was finally completed in 1938. The Colorado River Aqueduct built in 1939 connected the reservoir created by Parker Dam with city taps in Southern California. Other canals sent enormous volumes of water to irrigation projects in California's Imperial Valley and the Gila Valley of Arizona.

The increasingly heavy use of the Colorado's water over the past 80 years has created a highly precarious and contentious situation for all that depend on the river today. Phillip Fradkin, writing in his book *A River No More*,[2] described the Colorado as "the most used, most dramatic, and most highly litigated and politicized river in this country." The interstate water compact of 1922 set the stage for a litigious drama that continues to this day. That agreement, which sliced the Colorado's water pie into

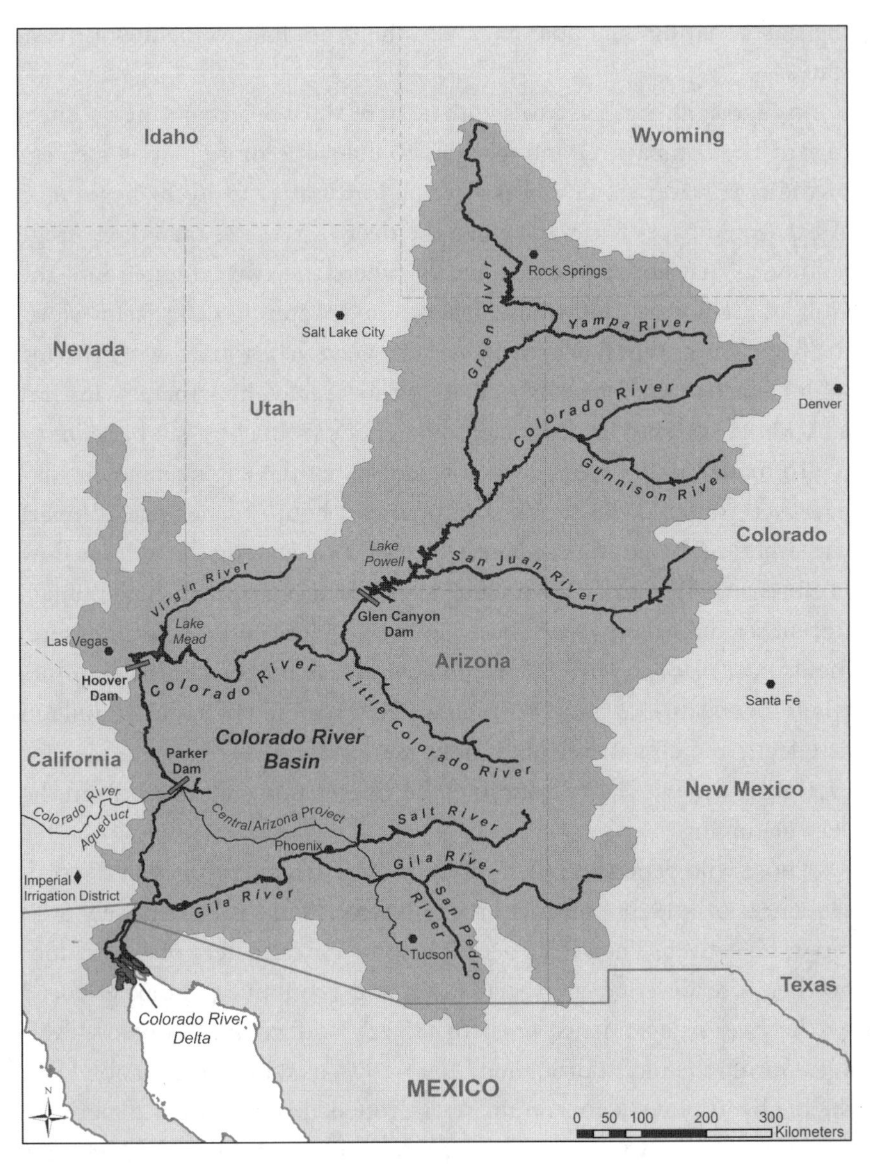

Figure 1-1. *Map of the Colorado River system.*

seven pieces for the states sharing the river, was based on an estimate that the river carried 21.6 billion cubic meters (17.5 million acre-feet[3]) of water each year on average. The compact allocated 19.7 billion cubic meters (16 million acre-feet) among the seven states, with roughly half of it going to the states sharing the upper river, and the other half to the downstream states.[4]

In hindsight, it is easy to see that there were defects in the architecture of the compact. Granting rights to consume more than 90 percent of the river's average flow was a recipe for disaster from the beginning: What would happen in years when the river carried less than the average volume? What about Mexico's needs, where the river emptied into the Gulf of California, nourishing fisheries and watering delta farms along the way? The compact acknowledged the need to negotiate water sharing with Mexico in the future, but did its authors really think that leaving just a trickle of water to its downstream neighbor would be a fair bargain?

To make matters worse, we now know that the engineers of the day had overestimated the river's average water bounty when the compact was drafted. The period used as the basis for calculating the average flow of the river—1905 to 1922—included periods of abnormally high rainfall. Recent scientific assessments based on a much longer period of measurements now place the river's average flow at least 15 percent lower, somewhere between 17.6 and 18.5 billion cubic meters (14.3 and 15 million acre-feet), and climate scientists now caution that the river is headed for even drier times.[5] The bottom line: the river was overallocated from the very beginning.

Those who depend on the Colorado River are still living with the consequences of a 1920s compact based on wrong data and exclusively utilitarian objectives. The sterile accounting of cubic meters of water does not begin to describe the social strain and economic pain experienced by the river's dependents when the river's performance is below average. During the recent drought of 1999–2003, electricity generation from the big hydropower dams on the river dropped by more than 20 percent, sending a shock wave through electricity bills across the southwestern United States.[6] Farmers watched their irrigation canals run dry and their crops wither. Lowered water levels in Lake Mead, the reservoir formed by Hoover Dam, left boat marinas high and dry, leading to 900,000 fewer tourist visits and lost revenues of $28 million, along with a loss of 680 jobs in the area.[7]

The most telling measure of the compact's failings is the fact that the river today runs completely dry before reaching the sea—not just in dry years, but in virtually every year. Many would assert that fully consuming a river's water is our manifest destiny, or the logical conclusion of using a precious resource to its fullest potential. But others see much more than cubic meters and kilowatts in a river.

In describing the Colorado River Compact in a *Los Angeles Times* editorial in 2012, water scientist and author Sandra Postel lamented, "All seven U.S. states in the basin were represented, but two voices were missing. One was that of Mexico. The other was the river itself."[8]

With every available crumb of the Colorado's water pie being consumed, the river has lost much of its once legendary ecological richness. In the river's delta, a place that the great American conservationist Aldo Leopold once described as a "wilderness of milk and honey" punctuated with deep green lagoons and marshes full of waterfowl and fish, now only a sunbaked, salt-caked barren wasteland remains. As the river ran dry, fisheries in the delta and the Gulf of California in Mexico were decimated, including a large population of totoaba (*Totoaba macdonaldi*), a fish that can grow to more than 100 kilograms (250 pounds). Any visitor to the desiccated delta today would find it very difficult to envision a behemoth fish like the totoaba swimming there. It would be like trying to imagine life on other planets.

The overallocated river has spawned social inequity as well. For more than a 1,000 years, an indigenous tribe of Cucapá—the "people of the river"—has relied on fishing and subsistence farming in the delta. When the river compact was being negotiated in 1922, no messenger was sent to the delta to invite the Cucapá to the bargaining table. When Mexico in 1944 wrestled with the United States for rights to the last of the river's dregs, Mexican officials were seeking water to grow asparagus and cotton in Mexicali, not fish or melons to feed the Cucapá. The river's natural bounty once supported more than 6,000 Cucapá. Fewer than 600 remain.[9]

History has in many ways validated Governor Moeur's anxieties of nearly a century ago: there are limits to what a river can give, and great care must be given to judgments about sharing water, because those decisions can bind like a knot and become difficult to untie. As the states sharing the river continue to negotiate its future, they should ask a question never asked as the Colorado River Compact was being negotiated 90 years ago: *Do we really want to take it all?*

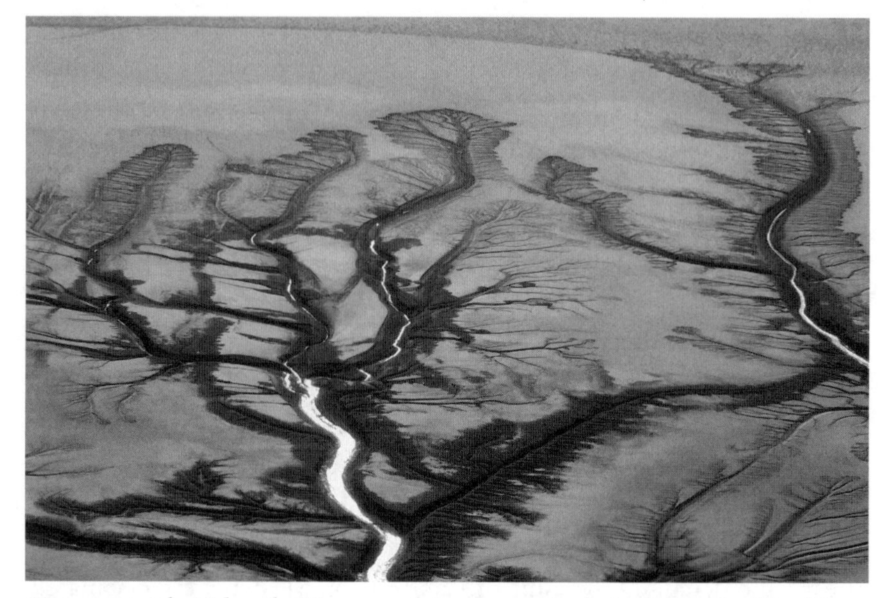

Figure 1-2. *The Colorado River now dries up completely in the sands of its delta, breaking its fluid connection with the Gulf of California in Mexico. Can a river still be a river when the water is all gone? (Photo by Pete McBride.)*

We possess the means to wring every last drop from the planet's rivers and lakes, or to suck its aquifers dry, but is that what we want to do? Or do we instead want to leave some water alone, enough to fuel the biological engines of our planet, or to serve as a hedge against dry times and an uncertain future, or simply to irrigate our souls with the intrinsic beauty of flowing water? Can a river still be a river when the water is all gone (fig. 1-2)?

Those questions are now increasingly being asked and answered for depleted rivers, lakes, and aquifers around the world. Australia's minister of water, Tony Burke, put it this way in explaining the goals of the 2012 Basin Plan for the Murray-Darling system: "Everybody needs to have healthy rivers. No one wins when our rivers die. And what's been happening for a long time now is that we've pulled so much water out of the rivers that they're living as though it's drought, ages before any drought actually arrives."[10]

When rivers and other freshwater ecosystems are maintained in healthy ecological condition, they provide myriad benefits and services to society. Natural vegetation such as floodplain forests and wetlands slow the flow of water, reducing its destructive force. As the water slows, it is cleansed by biological processes and can recharge groundwater aquifers. The freshwater moving through a river ecosystem to coastal areas maintains proper salt balances and carries nutrients to estuaries, enabling fish, shellfish, and other animals to breed and grow. The sediment carried by rivers forms and replenishes coastal beaches and barrier islands that buffer the shoreline against ocean storms. Nature does this invaluable work for free, but only if we leave enough water to nature to allow it to function properly.

We did not adequately understand these natural processes and their importance to our well-being in 1922 when the entirety of the Colorado River was allocated for human use, nor have these natural ecosystem services been adequately considered as so many other rivers, lakes, and aquifers have been depleted. Scientists are now much better able to quantify how much water must be left to sustain nature's health. Fortunately, consideration of those ecosystem needs is finally beginning to appear in water-sharing plans and agreements around the world.[11]

A Lifetime of Worrying over Water

Colorado River water ran through my own veins for the first two decades of my life, which I suspect left me imprinted with an awareness of water scarcity. My father became smitten with San Diego while spending 4 years there during his service in the U.S. Navy in the early 1950s. When he left the navy in 1955, he went back home to Texas, married my mother, drove her to San Diego on their honeymoon, and then welcomed me into the world a year later. My parents had escaped Texas in the nick of time. The worst drought in history was wracking the state, killing hundreds of thousands of cows and leaving stunted corn rotting in the fields. That drought drove half of all Texas ranchers and farmers into bankruptcy, and the rest into near madness.

My family rode the wave of optimism and population growth in San Diego that was unleashed by the redirected flow of Colorado River water into Southern California in the 1940s. The city grew by a half million new residents by my tenth birthday, severely straining the limits of the city's entitlement to the Colorado's water. A drought during my teenage years inspired a popular bumper sticker slogan urging us to "Save Water—Shower with a Friend." The state of California eased its water strain by building the State Water Project in 1978, reaching 800 kilometers (500 miles) into Northern California's rivers to bring water to southern coastal cities. By the time I left for graduate school in the 1980s, San Diego was importing 95 percent of its water supply from far-distant rivers. The city has never stopped looking for more water.

I have never stopped thinking about droughts and water shortages. I remember in high school pondering the thought that if I were to become a water expert, I would have job security for life. I have spent the past 25 years on that path through my work with The Nature Conservancy, which has provided me the opportunity to travel the world working on water solutions to benefit people and nature. I have witnessed firsthand the consequences of water shortages, providing me with some insight. Throughout those years and travels, I kept asking the same questions I first asked when I was learning of the Colorado River's limits:

- Who was experiencing water shortages, and where?
- What happens to people and other species when water shortages occur?
- Why do communities and countries run short on water?
- Is there some way to avoid shortages, or overcome them once afflicted?

In this book, I will share some of what I have learned and offer my own evolving answers to these questions. In the remainder of this chapter, I will provide an overview of the places where water scarcity is happening and begin describing the impacts of water shortages around the world. That will leave the remainder of the book to explain what causes water shortages and what we might do about them.

Water Scarcity in the World

In 1985, Boutros Boutros-Ghali—who would become secretary-general of the United Nations 7 years later—warned that "the next war in the Middle East will be fought over water, not politics." It was a warning shot heard around the world, leaving many countries to ponder their own water futures. More recent water warnings have not been limited to the Middle East. Kofi Annan, Boutros-Ghali's successor at the UN, cautioned in 2001 that "fierce competition for fresh water may well become a source for conflict and wars in the future." He was in turn succeeded at the UN by Ban Ki-moon, who expressed concern in 2008 that many conflicts around the world were being fuelled or exacerbated by water shortages. In 2013, Ban Ki-moon went a step further, warning that the world was on course to run out of freshwater unless greater efforts were made to secure water supplies.

The 1934 conflagration over Parker Dam on the Colorado River marked the last time in U.S. history that a state took up arms against another state, but not the last water fight to take place beyond the United States. In fact, as suggested by the UN's repeated cautionary warnings, water conflicts have grown in number around the world and intensified over time as water supplies have become increasingly strained.

Aaron Wolf and his colleagues at Oregon State University have been carefully documenting the history of water skirmishes around the world. Aaron points out that the last true all-out water war among nations took place more than 4,500 years ago, between the city-states of Umma and Lagash in the Tigris-Euphrates river basin, in the area of present-day Iraq. However, the absence of outright military action belies the potency of water scarcity as an undercurrent in political conflict and social unrest. The Six-Day War in the Middle East in 1967, for example, was in part sparked by tension over a Syrian project to divert the Jordan River, and water remains a divisive issue between Israel and its neighbors to this day. Because of long-running contention and lack of coordinated management of the Jordan River, it has shriveled to a fraction of its original size, and the water level in the Dead Sea—into which the river flows—has been dropping by more than a meter each year. This drying of precious water sources, including heavy depletion of the region's aquifers, greatly exacerbates political tension in the area.

Water itself has been used as a weapon. In the early 1990s, Saddam Hussein punished the Marsh Arabs, who are Shi'a Muslims in southern Iraq, for their insurrection against his regime, by using dams to shut off the flow of the Tigris and Euphrates Rivers into the Mesopotamian Marshes. The marshes, once the third-largest wetland area in the world, have been occupied for more than 5,000 years. But without adequate water flow, they quickly withered to a tiny fraction of their original size. The Marsh Arabs, whose population was estimated at 250,000 in 1991, dropped precipitously to less than 40,000.[12]

Much more recently, in October 2012, more than 5,000 farmers and activists from the state of Karnataka in India attempted to seize control of the Krishna Raja Sagar Dam on the Cauvery River, in an effort to shut off water releases to the downstream state of Tamil Nadu. Unlike Saddam Hussein, the people of Karnataka were not doing this to punish their downstream neighbors, although more than 120 years of dispute over the Cauvery's water has certainly brewed a good deal of hatred between these Indian states. Instead, they were acting out of desperation. During droughts such as that in 2012, there simply is not enough water in the Cauvery to irrigate their farms. To them, water hoarding is the only alternative to going broke or starving to death.

In the past couple of decades, I have listened to many heated debates among neighbors, communities, and countries over water shortages. I have seen the tension, hostility, and suffering that comes with water scarcity, as well as its devastating impacts on wildlife and ecosystem functioning, from the Klamath River in Oregon to the Apalachicola River that flows through Georgia, Alabama, and Florida in the United States; from the Tana River in Kenya to the Zambezi River in Mozambique; from the Yaqui River in Mexico to the Yangtze River in China. In my own hometown of Charlottesville, Virginia, a severe drought in 2002 nearly drained the community's water supply, falsifying the long-held belief that water shortages threatened only the western half of our country.

The more I learned about these places and conflicts, the more I wanted to dive deeper into understanding the incidence and spread of water scarcity across the globe. I wanted to better understand how water was being used in those places, what had led to shortages, and what could be done to resolve conflicts in water-strapped places.

I began working with Martina Flörke at the University of Kassel in Germany and Kate Brauman at the University of Minnesota to develop some new global maps of water scarcity. Martina has been a key contributor to the development of a global hydrology model called WaterGAP. She is one of a growing number of researchers around the world who are developing computer models to assess the status of our planet's water sources. Most of the global water modeling groups are based in academic institutions such as the University of Kassel and University of Frankfurt in Germany, Utrecht University and the University of Twente in the Netherlands, and the City College of New York.

Using supercomputers to perform the millions of calculations in their models, these global water modelers use the best available estimates of rainfall, snowfall, evaporation, water use, and many other factors to gain an understanding of water conditions around the globe. They calibrate their models by comparing their model results with on-the-ground data collected in tens of thousands of monitoring stations where river flow or groundwater levels are being measured. They have recently even begun to use water measurements taken from satellites that provide estimates of changes in the water volumes of lakes, glaciers, ice caps, and aquifers.[13]

While many different approaches and indicators have been used to characterize or quantify water scarcity, Martina, Kate, and I came up with a very simple "water depletion" index that we use to depict the water scarcity status of each water basin.[14] Using outputs from the WaterGAP model, we have estimated how much of the average water flow is left in more than 11,000 water basins after being used in cities, industries, and farms. We have compared that remaining volume of water against the volume that we would expect to find if nobody was using the water. In essence, we have tried to depict how much water—on a percentage basis—is regularly being depleted from each water source by human use. The map in figure 1-3 is one product of our efforts.

We developed this index of water depletion because it is clear to us, as so well illustrated by the Colorado River, that as local water sources are increasingly depleted, the vulnerability of local communities and aquatic ecosystems to severe impacts from water shortages goes up accordingly. Our assessment has revealed that—unlike the Colorado River—few places are experiencing water shortages on a continual basis. Instead, water shortages tend to be episodic in nature, emerging during dry seasons or during droughts. It can be very difficult or expensive for entire communities of water users to reduce their consumption quickly, or to a large

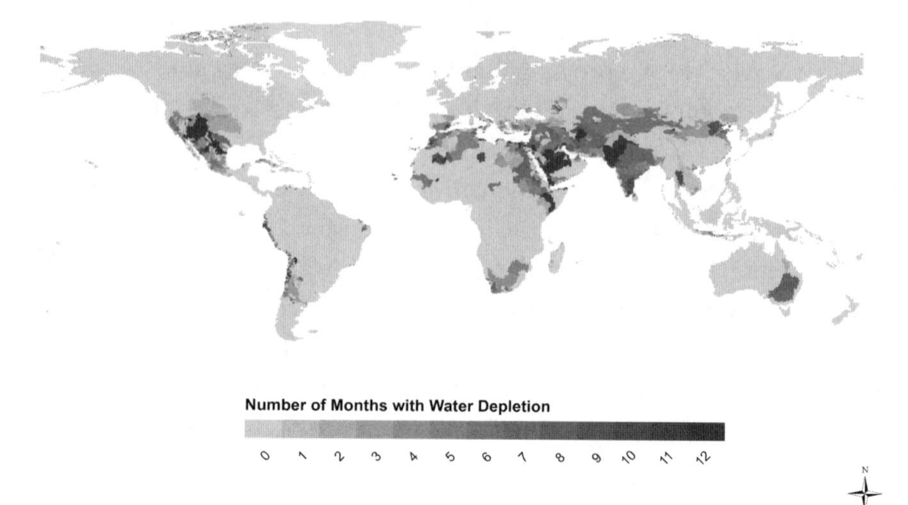

Figure 1-3. *This global map of water scarcity highlights places where more than half of the average water flow is being depleted during one or more months of the year. (Produced using the WaterGAP model developed at the University of Kassel in Germany.)*

degree, during a dry period. For that reason, when water users get into the habit of consuming a large portion of a water source's yield on a regular basis, they are poised for disaster when those dry periods eventually arrive.

This vulnerability to episodic water shortages can be illustrated with the Brazos River in Texas. Figure 1-4 illustrates the degree to which the river was depleted during each month over a typical 2-year period. On average, about half of the Brazos is depleted, but heavy use of the river causes it to dry to very low levels in months or seasons lacking rainfall, placing all water users in jeopardy.

One of those users is Dow Chemical's manufacturing plant in Freeport, Texas, located near the river's mouth. In November 2012, fearing that its plant—one of the largest chemical manufacturing plants in the world, with more than 8,000 employees—would run out of water if drought conditions persisted, the company petitioned the state to shut off all water users with lower-priority water rights, to ensure that the company's water entitlement would be satisfied. Because such an action would have left more than 700 farmers without water, the Texas Farm Bureau quickly filed a lawsuit against the state to prevent the cutoffs.[15]

Fortunately, the tension was relieved in early 2013 as rainfall refreshed the river's flow, but without substantial reduction in water use along the Brazos, this reprieve is likely to be only a temporary one.

Our map in figure 1-3 illustrates the fact that water scarcity is not happening everywhere. We found that less than one-quarter of all water sources are experiencing some months with water depletions of more than 50 percent. Of great concern, however, is the fact that half of the world's cities with populations greater than 100,000 are situated in those water-stressed places. Many of those cities are having great difficulty in securing the water supplies needed to support their growing populations.

The hot spots on our map correspond to rivers that are regularly drying up completely, such as the Yellow and Tarim Rivers of China, the

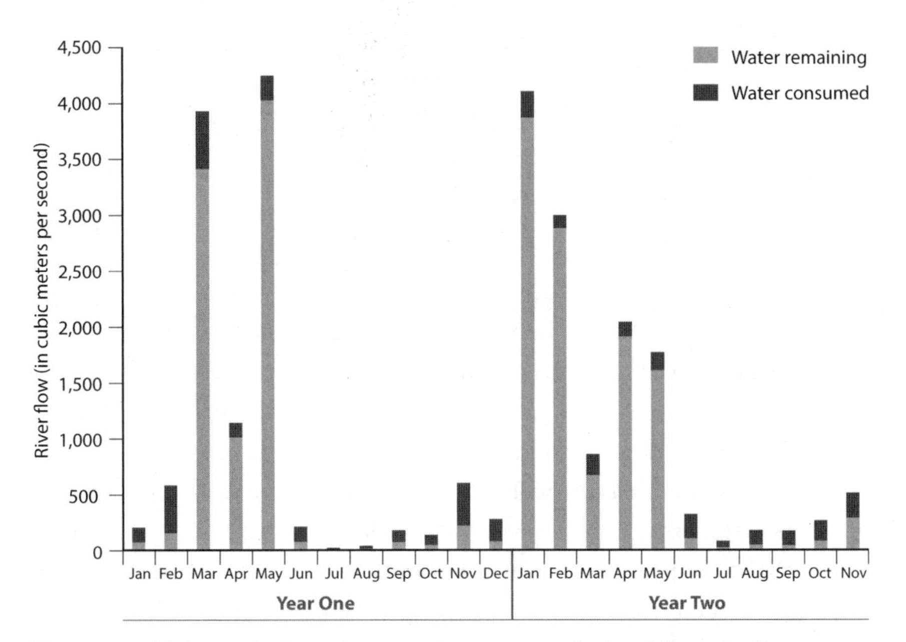

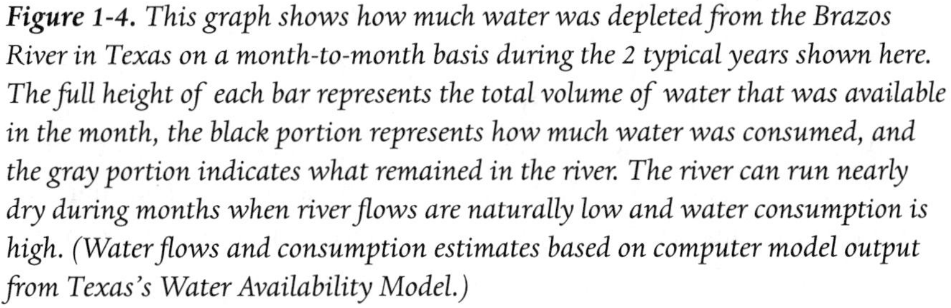

Figure 1-4. *This graph shows how much water was depleted from the Brazos River in Texas on a month-to-month basis during the 2 typical years shown here. The full height of each bar represents the total volume of water that was available in the month, the black portion represents how much water was consumed, and the gray portion indicates what remained in the river. The river can run nearly dry during months when river flows are naturally low and water consumption is high. (Water flows and consumption estimates based on computer model output from Texas's Water Availability Model.)*

Colorado River and Rio Grande of the United States, the Indus of Pakistan, and the Murray-Darling system of Australia. Other water-scarcity hot spots include the depleted High Plains Aquifer in the United States, the North Arabian Aquifer of Saudi Arabia, and the desiccated Aral Sea in Kazakhstan and Uzbekistan. Table 1-1 provides a list of some of the most depleted water sources according to our estimates. We have also documented how the water is used in water-scarce places; more than 90 percent of all water depletions go to irrigated agriculture.

The Pain of Water Scarcity

The World Economic Forum has now placed water supply crises near the top of its list of global risks, on the basis of potential impact and probability.[16] Abundant evidence of water hardship can be found by searching the Internet for water news related to any of the water sources listed in table 1-1. Together with my students at the University of Virginia, I have been compiling a global database of the economic and other impacts caused by water shortages.[17] Whenever we find a news article, journal paper, or website that discusses the impacts of water shortages somewhere in the world, we make note of it in our database. Similarly, Aaron Wolf and his colleagues at Oregon State University have been building the Transboundary Freshwater Dispute Database[18] to identify places where water conflicts have erupted or where they are being resolved, such as through water treaties among countries that share the same river or aquifer.

Some generalities can be drawn from the many case histories of water scarcity accumulating around the world:

The economic impacts of running out of water can be devastating. The state of Texas lost nearly $9 billion of revenue during a drought in 2011, with much of that loss on farms that could not be irrigated, because of water shortages. China loses $15 billion each year because of groundwater depletion, and another $24 billion in lost water availability due to pollution.[19]

The cost of securing additional water can be very expensive. When communities exhaust their local water sources, additional water will have to be obtained from elsewhere. Construction of a pipeline might be necessary, to bring in water from other places, or a desalination plant might be needed so that seawater or brackish (salty) groundwater can be turned into freshwater. These options come with heavy up-front construction

Table 1-1 The World's Most Depleted Freshwater Sources

Aral Sea, Kazakhstan/Uzbekistan	Persian Aquifer, Iran
Krishna River, India	Dead Sea, Jordan/Israel
Armeria River, Mexico	Rio Grande / Rio Bravo, USA/ Mexico
Loa River, Chile	Doring River, South Africa
Brazos River, USA	Sacramento River, USA
Lower Indus Aquifer, India/Pakistan	Fuerte River, Mexico
Cauvery River, India	San Joaquin River, USA
Mahi River, India	Ganges River, India/Bangladesh
Central Valley Aquifer, USA	Santiago River, Mexico
Murray-Darling Basin, Australia	Godavari River, India
Chao Phraya, Thailand	Shebelle River, Ethiopia/Somalia
Narmada River, India	Great Salt Lake, USA
Chira River, Ecuador & Peru	Tapti River, India
Nile Delta Aquifer, Egypt	High Plains Aquifer, USA
Colorado River, western USA	Upper Ganges Aquifer, India/ Pakistan
North Arabian Aquifer, Saudi Arabia	Huasco River, Chile
Colorado River (Texas), USA	Western Mexico Aquifer, Mexico
North China Plain Aquifer, China	Indus River, Pakistan/India
Colorado River, western USA	Yonding River, China
Penner River, India	
Conception River, Mexico	

costs and hefty, never-ending electricity costs for pumping and treating the water.

Water shortages impair the functioning of both gray (man-made) and green (natural) infrastructure systems. Power plants and hydropower dams cannot operate at full capacity, river navigation and barge transport come to a halt, rivers can no longer sufficiently dilute and assimilate the wastes they receive, and estuaries stop producing fish and shellfish because there is not enough freshwater inflow to maintain proper salt concentrations. Water depletion is a major cause of aquatic species imperilment.[20]

Human lives and livelihoods are compromised. Water shortages can be deadly for many poor people living in developing regions of the world who cannot readily access clean drinking water supplies. When local water sources are dried up or polluted, many—usually women and children—are forced to walk long distances to reach other water, taking a heavy toll on their health and their ability to participate in other chores or go to school. The decline in the ability of many poor families to grow their own food due to water shortages is leading to mass out-migration in many water-scarce regions. In the worst cases it is leading to bankruptcy, divorce, suicide, and fractured communities.

The quality of our lives is diminished. Everything that we value in flowing or ponded water, including recreating in it, relishing its aesthetic qualities, observing the other species it supports, and worshipping it through spiritual practice, is lost when a river or lake dries up.

Learning from the Past to Build a Better Water Future

The calamity of water scarcity in the Colorado River can in part be blamed on a lack of accurate information—the data available to scientists when the Colorado's waters were allocated among the states were inadequate for understanding the variable nature of the river's water bounty. Today, we are much better able to estimate and understand how much water may be available in our water sources during average years, dry years, and wet years. The question is, Will we use that information to better manage water in a way that sustains what we value, gives us what we need, and enables us to better avoid the perils of scarcity?

Hundreds of books and thousands of technical papers have been written on the subject of water management, and yet so many communities continue to crash into the wall of scarcity. We urgently need to design, experiment with, and give life to some fundamentally new forms of water democracy. The twentieth century taught us that top-down, state-run technocracies are simply not willing to or cannot properly allocate, monitor, and govern water in a way that would forestall scarcity. To avert scarcity going forward, I believe that we will need to enable and empower more localized decision-making and management processes that can be rightsized to the particular needs, uses, economics, and cultures associated with the sharing of water sources. Ultimately, effective water management will require both technical capacity and appropriate engagement of water users and other local interests.

Empowering local communities of water users will require that we overcome pervasive water illiteracy. The stark reality is that most people alive today could not begin to sketch the global water cycle, do not know how the water sources they depend upon are being used and by whom, and do not even know where their water comes from. Lacking such knowledge, they cannot possibly contribute to any sort of citizen-centered water democracy in any meaningful and productive way.

The World Economic Forum, in its global assessment of water scarcity, put it this way: "We are now on the verge of water bankruptcy in many places with no way of paying the debt back."[21] This allusion to financial accounts is quite apt. As discussed in the next chapter, solving the problem of water shortages begins by learning how to balance our water budgets.

CHAPTER TWO

Taking Stock of Our Water Budgets

A few years ago, I was asked to join a panel of speakers at a national conference on water scarcity. One of my fellow panelists was a farmer from Florida. After he listened to many presentations by others, the farmer was obviously anxious to get in a word. He confidently proclaimed that there is no such thing as water scarcity. He asserted that concepts like "water depletion" are fallacies: "We only borrow water for a short while, and then it all comes back in time."

It took me a while to digest and comprehend his meaning, but I soon realized that rather than talking about any particular local scenario, what he was referring to was the global water cycle. In grade school, we are taught that the water in the ocean evaporates and forms clouds; clouds then release their moisture onto land, where the water then either evaporates, soaks into soils, or gathers in lakes or flows downstream through rivers toward the ocean. On its way, we consume some of this water, and when the remainder eventually returns to the ocean, the cycle is complete. When water is viewed in this way, the farmer is absolutely correct: We borrow it, and it comes back. In fact, our planet has not lost any freshwater in millennia. It just keeps recirculating in the planetary cycle, over and over again.

Taking this whole-Earth perspective one step further, the farmer is also correct in stating that we do not have a water shortage—at least not a *global* shortage. The whole human enterprise—cities, industry, agriculture—is currently using only 12 percent of all of the water that flows into and through the planet's rivers, lakes, and aquifers on a continual basis.[1] And all of the water we use eventually ends up back in the rivers, the sky, or the ocean.

But as the people who live near the heavily depleted Colorado, Jordan, and Cauvery Rivers will tell you, it doesn't matter that there is plentiful water flowing down the Amazon, Yukon, or Congo. The water in those rivers can't be used to resolve *their* water shortages. They can't afford to bring it to Arizona, the Middle East, or India by ship or pipeline, even if they could secure legal rights to it. Human civilization and the world's great cities grew up along the banks of large rivers and lakes for a very basic reason: People need freshwater daily, and within easy or affordable reach. In that sense, water scarcity can be defined as a condition that occurs when there is insufficient water available at reasonable cost to fulfill human needs and to sustain the health of freshwater ecosystems. That will necessarily include leaving some of it alone, to flow in a river or remain in a lake or aquifer.

Water scarcity is not global in its physical expression—instead, it is highly localized. Most places in the world are not experiencing serious or regular water shortages at this time. However, the communities using the waters of the Colorado, Jordan, Cauvery, and hundreds of other stressed rivers and aquifers around the world are in serious trouble for two simple reasons: (1) they have been consuming their available water faster than it can be regularly replenished with rain and snow, and (2) they lack sufficient restraint or regulatory controls, or are too overwhelmed with the struggles of living day to day, to keep this from happening.[2] In this chapter I'll address the first part of this problem, and I'll begin discussing options for water governance in chapter 4.

Managing the Water Account

Managing water sources for long-term sustainability will require an accurate accounting and understanding of how much water is available and how it is being used. Your personal bank account provides a useful analogy for understanding water scarcity. As with water in the global cycle, it doesn't matter to you that there are trillions of dollars moving through the global economy. You only care about how much is moving through your own bank account.

The water analog to your bank account is a watershed or an aquifer (see fig. 2-1 and 2-2). The water that you rely on in your daily life comes from these sources. When rain or snow falls from the sky and lands on

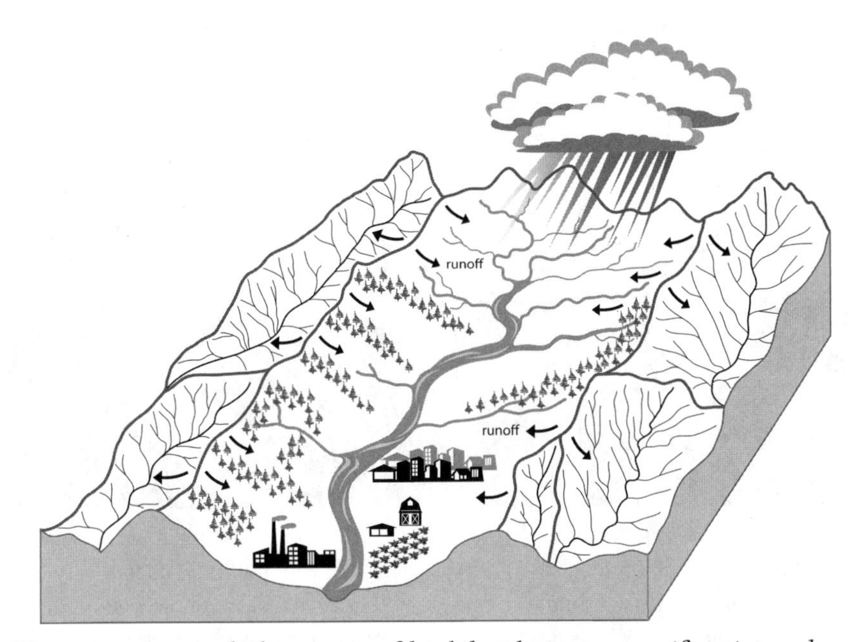

Figure 2-1. *A watershed is an area of land that drains to a specific point, such as to a lake, an ocean, or a location along a river. For instance, as you stand on the bank of a river, the watershed for your location is composed of the landscape area that drains to that point. Water from rain or snow falling anywhere in a watershed moves downhill (as runoff), and some of it eventually reaches your location. Along the way to your location, some water evaporates, and some is used by vegetation or for human purposes. Six different watersheds are delineated in this picture.*

the earth, some water will percolate deep into the ground and end up in an underground aquifer. Other water will run off the land and into a river or lake. The area of land draining to a particular river or lake forms that water body's "watershed." As water collects in a watershed or aquifer, it becomes available for use. It's your water account.

To manage your bank account in a sustainable manner, you should always avoid spending more than you deposit. The same is true for watersheds and aquifers. To do this well, you must keep an accurate accounting of all deposits and expenditures, and you must limit your spending within a budget.

Unlike your personal bank account, though, your water account is shared by many other people. That means that managing your water account will require rules to guide everyone's use of the water account.

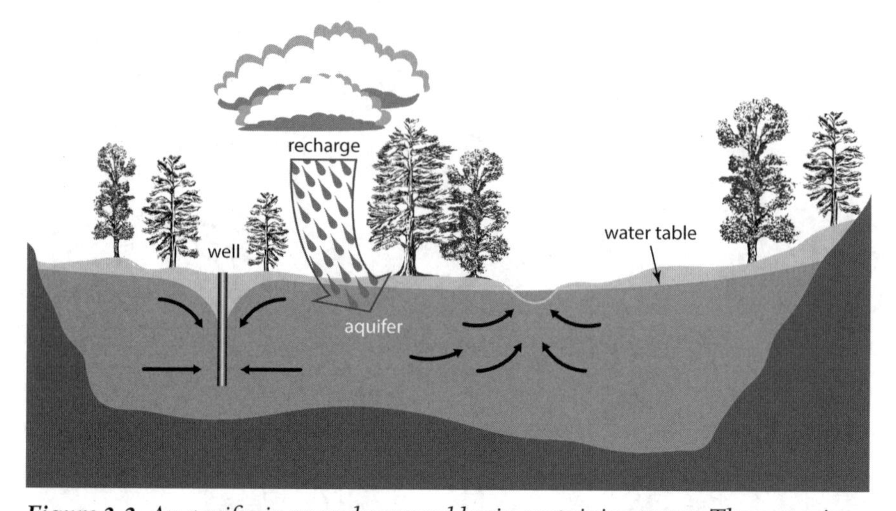

Figure 2-2. *An aquifer is an underground basin containing water. The water in an aquifer can be extracted using groundwater wells and pumps. The water in aquifers originates from rain or melting snow that percolates into the ground to recharge the aquifers. Water has been accumulating in some aquifers for thousands or millions of years. In some geologic settings, such as a limestone formation, the water will collect in subterranean caverns and channels. More common, however, is for the water to soak into buried sediments or porous rock—think of a swimming pool filled with sand, with water saturating the sand to a certain level. The water level in an aquifer is called the water table. When water is consumed from the aquifer faster than it is being recharged, the water table will be lowered, sometimes to the point that wells can no longer reach the water or it becomes too costly to pump the water from great depths.*

As will be discussed further in chapter 4, those rules can be set unilaterally by your government, or they can be agreed upon through community dialogue, or some combination of approaches can be used. Similarly, enforcement of the rules can be done in a variety of ways.

The Vocabulary of Water Budgets

Before I illustrate additional water budget principles with real-world examples, definitions of some key terms that are used in the language of hydrology and water budgets are important. These terms are illustrated in figure 2-3.

Rain and snow are natural deposits into a water account, and they are referred to as the natural or renewable "water supply."[3] Extraction of water from a water source, such as a river or lake or aquifer, is called a "water withdrawal." After withdrawn water is used on a farm or in a city or industry, some portion of that water runs off the farm or down the drain and reenters the original water source as "return flow." The portion of used water that is *not returned to the original source* is considered a "consumptive use," that is, it is lost or depleted from the water account.

When constructing a water budget for a particular water source, it is very important to pay attention to water being moved between water sources. Many cities withdraw their water from one freshwater source but return it to another freshwater source or to the ocean. All of the water that is not returned to the original source is considered to be consumptively used (depleted) with respect to the original source. This is

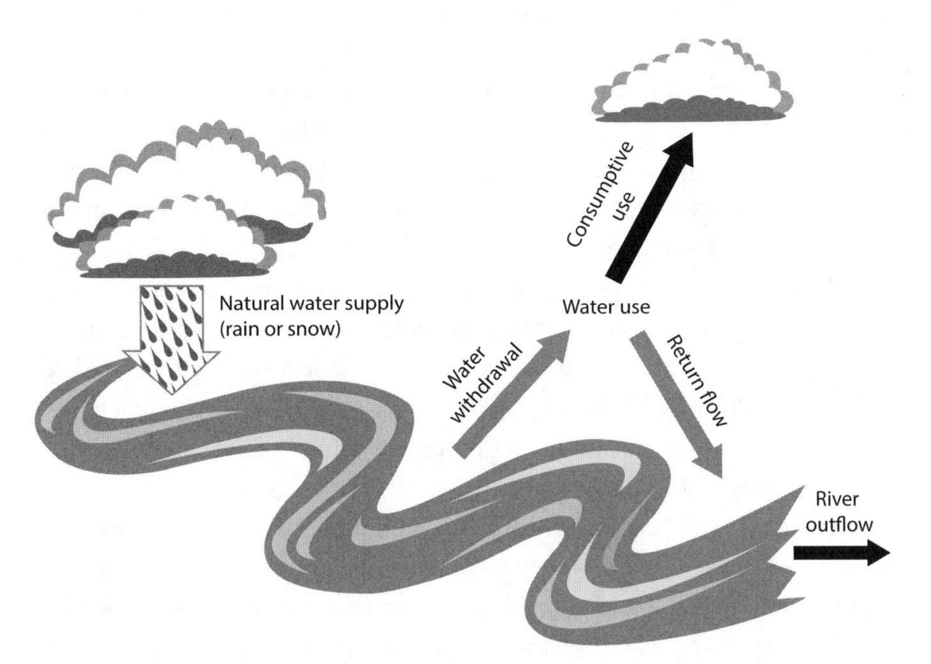

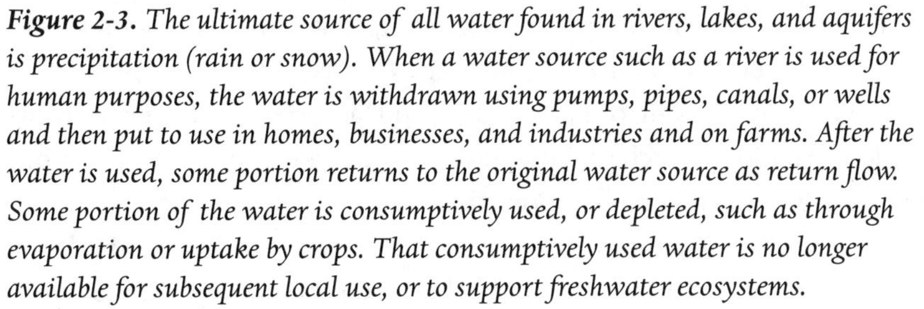
Figure 2-3. *The ultimate source of all water found in rivers, lakes, and aquifers is precipitation (rain or snow). When a water source such as a river is used for human purposes, the water is withdrawn using pumps, pipes, canals, or wells and then put to use in homes, businesses, and industries and on farms. After the water is used, some portion returns to the original water source as return flow. Some portion of the water is consumptively used, or depleted, such as through evaporation or uptake by crops. That consumptively used water is no longer available for subsequent local use, or to support freshwater ecosystems.*

particularly true for groundwater use—water is seldom returned to an aquifer after it is used. Water budgets for watersheds and aquifers should be calculated independently—even when they are in the same geographic location—but with an accounting for any exchanges between the two.

In the calculation of a water budget, the water withdrawals, uses, return flows, and consumptive uses are usually tracked according to various subaccounts (i.e., types of use). While many different subaccounts can be defined, the four most commonly used categories include irrigated agriculture, domestic (home or business) use, industrial use, and electricity generation. These different types of water use will consumptively use differing portions of the volume of water that is withdrawn.

The consumptive use of water for agriculture, for example, is typically 50 to 60 percent of the volume of water that is withdrawn; the remaining 40 to 50 percent makes its way back to a water source as return flow (table 2-1; also see fig. 2-3). Consumptive losses of water in agriculture are caused by water evaporating from the soil, plant leaves, or canals; water being taken up by plants and then transpired (evaporated); or water percolating deep into the ground and becoming unavailable for subsequent use. Compared with flood irrigation, in which a farm field is completely inundated with water, highly efficient irrigation practices such as drip irrigation apply water only in the immediate vicinity of the plants, using plastic pipes and tubes. As a result, less water is lost to evaporation, and so less water needs to be withdrawn from a water source in the first place. But virtually all of the water that is applied using highly efficient irrigation practices will be consumptively used.

About 10 to 30 percent of the water withdrawn and used for domestic purposes will be consumptively lost on average, but those losses can get as high as 50 to 70 percent of withdrawn water if a lot of water is used for outdoor landscape watering. Industries consumptively use 5 to 20 percent of the water they withdraw, and most energy facilities, such as a power plant that generates electricity, consumptively use only 2 to 5 percent of withdrawn water. Think of farms, cities, and factories as huge water recirculation machines—each of them takes in a certain volume of water but then usually returns some portion after use.

The water returning to a water source after use will often be heavily contaminated with nutrients or other chemicals, or overloaded with sediments or organic material. This can make the returned water unsuitable for subsequent use, and unhealthy for fish and other aquatic species. Technically, this water is not consumptively used or lost from the local water account if it is feasible to use it again, but it can be very expensive to clean it up. This is a very serious problem in some developing regions, where as much as 90 percent of the water returned after use has received no cleansing whatsoever.

To understand how a water source is being depleted, you will need to account for both water withdrawals and consumptive uses. Once you understand how the available water supply is being reduced by consumptive use, you can design effective strategies for either supplementing the remaining water

Table 2-1 A Comparison of Water Withdrawals and Consumptive Use

United States Water Use				
	Volume of Withdrawals (BCM*)	Percent of Total Withdrawals	Volume of Consumptive Use (BCM)	Percent of Total Consumptive Use
Domestic Use	57	10	33	25
Industrial & Mining Use	57	10	5	4
Electricity Generation	278	49	7	5
Irrigated Agriculture & Livestock Use	176	31	88	66
Totals	568	100	133	100
Global Water Use				
Domestic Use	380	10	42	4
Industrial & Mining Use**	780	21	38	4
Irrigated Agriculture & Livestock Use	2,600	69	945	92
Totals	3,760	100	1,025	100

* BCM = billion cubic meters.
** includes electricity generation.

Sources used in preparing this table include the U.S. Geological Survey, U.S. Department of Agriculture, and UNESCO-IHE Institute for Water Education and Electric Power Research Institute. For detail see the Acknowledgments.

by using other sources, or reducing the volume of consumptive use, as discussed further in chapter 3. You will also need to look at the additional volume of water that people want or need to withdraw and assess whether sufficient water remains to satisfy the necessary withdrawals.

For example, let us assume that 100 units of water are flowing in a river. Farmers along the river consumptively use 60 units of that water, leaving only 40 units flowing downstream. Of those 40 units, 30 units need to be left in the river to support fish and other aquatic life, as well as to dilute any wastes (i.e., from factories or farms) that are being carried by the river. Thus, only 10 units remain for additional human uses. A power plant in the lower river needs to withdraw and use 20 units for plant cooling. Even though the power plant would consumptively lose only 2 units of withdrawn water to evaporation, there is not enough water left in the river to satisfy its withdrawal needs (i.e., it needs 20 units for cooling but only 10 are available).

Figure 2-4 illustrates the proper way to view water withdrawals and consumptive use when trying to address water scarcity or to resolve recurring water shortages in a particular location such as a farm or city. You must first look upstream, or look to other users of your aquifer, and consider how water is being consumptively used. That will help you understand why water is sometimes or always scarce at the location of your use. Then take a look at how much water you need to withdraw and use, and consider whether—or how often—you might not have enough water to meet your needs. When there is insufficient water available at your location to meet your needs, you should next consider which of the following three options is most feasible or desirable: (1) Can you work with those upstream or other aquifer users to reduce the amount of water they are consumptively using? (2) Can you reduce the amount of water you need to withdraw and use at your location? (3) Can you somehow supplement the amount of water available to you, such as by bringing water in from outside your local watershed or aquifer?

As shown by Table 2-1, the use of water for domestic, commercial, industrial, or mining purposes combined usually accounts for only a minor fraction of all water withdrawals from most water sources, and an even smaller fraction of all water consumptively used. While electricity-generating facilities often withdraw and use large volumes of water, they

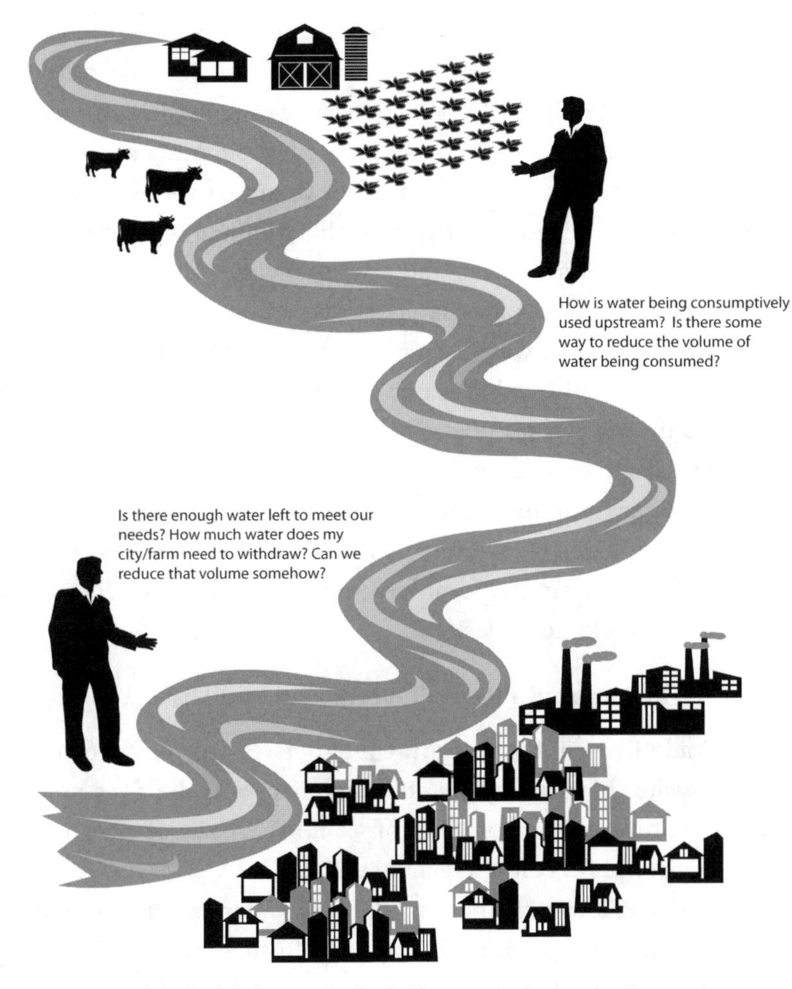

Figure 2-4. *To gain an understanding of why water shortages are occurring at your location, it is important to first look at how water is being consumptively used upstream and to assess whether those consumptive losses can be reduced. It is also important to look at how water is being used at your location of water withdrawal and to assess whether those needs can be reduced.*

represent only a tiny fraction (about 3–5 percent) of all consumptive use. In most countries, irrigated agriculture withdraws and consumptively uses the greatest volumes of water, by far.

These numbers shed light on the rapidly growing tensions between urban and rural users of water. Even though urban water uses (for domestic, commercial, and industrial purposes) account for only a minor share of total water use in most water-stressed places, urban water use is often growing much faster than agricultural use. The need for more water in cities and industries conflicts with the reality that agriculture is already consumptively using most of the renewable water supply. One of the options available to cities or industries wanting to access more water is to find ways to work with farmers to share limited water supplies. Ultimately, the economic, social, and environmental risks associated with water scarcity cannot be adequately addressed in many places without reducing the volume of water being consumptively used in agriculture.

While the U.S. and global water data presented in table 2-1 provide insight into the general purposes for which water is withdrawn and consumptively used, the underlying causes of your water shortages can only be understood by examining the particular situation in your local watershed or aquifer. Effective strategies for alleviating water shortages can be designed using accurate, local information, as illustrated here with a couple of real-world examples.

The Colorado River: Overdrawn and Dried Up

The Colorado River in the western United States provides a telling illustration of water bankruptcy. As noted in chapter 1, the river no longer reaches its natural delta at the Gulf of California, because all of its water is consumptively used before reaching the sea.

Nearly 170 billion cubic meters (BCM) of water, on average, fall from the sky as rain or snow into the Colorado River's watershed each year. Only about 12 percent of that precipitation—about 20 BCM—makes its way into the Colorado River and its tributary streams; the rest evaporates or is used up by plants before getting to the river system.[4] The water users dependent on the Colorado River—farms, cities, industries, mines, power plants—withdraw more than 30 BCM from the river on average. You might ask, How can more water be withdrawn from the river than is available?

The disparity between water supply and water withdrawals in this case can be largely explained by return flows. Much of the water withdrawn from the river is returned after use and thereby becomes available for other users downstream. For that reason, the river is able to support much more withdrawal and use of water than would appear to be available. In many watersheds, some portion of the return flow passes downriver to the river's mouth, but that's not the case in the Colorado, where every drop is consumptively used before reaching the delta.

This Colorado River example very well illustrates why it is so important to pay attention to consumptive uses of water, in addition to water withdrawals, when assessing water scarcity. Unfortunately, most tallies of water use provide only estimates of water withdrawals and not consumptive use.[5] Those water withdrawal estimates can give a very misleading picture of water scarcity because they imply that much more water is being depleted from our freshwater sources than actually is. In the Colorado River watershed, looking only at water withdrawal estimates would

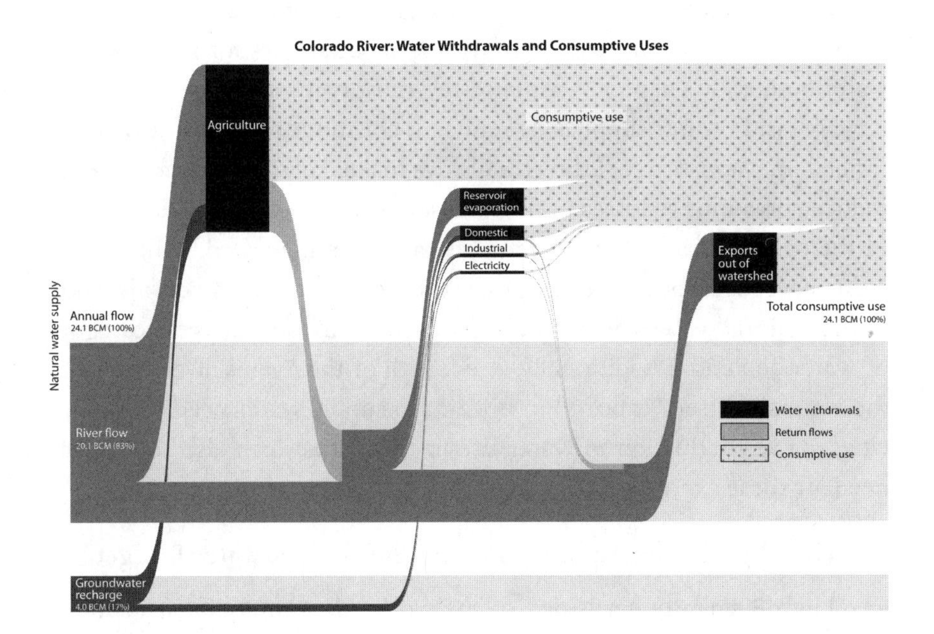

Figure 2-5. *This diagram depicts the natural water supply, use, return flow, and consumptive loss of water in the Colorado River watershed. The far left side of the diagram indicates the average volume of water in the river and underground aquifers that is renewed by rain and snow annually. This renewable water is then withdrawn and used for various purposes, with some portion of the withdrawn water being consumptively used (depleted) and some returning to the river after use. Before reaching the river's delta in Mexico, all of the water has been consumed. (Adapted from the original illustration by Jason Pearson of TruthStudio, based on data from the U.S. Geological Survey and U.S. Bureau of Reclamation.)*

suggest that 50 percent more water is taken from the river than is actually available, which is of course impossible.

Figure 2-5 diagrams the volume of water that is withdrawn, consumptively used, and returned in the Colorado River's largest water-use categories. Typical of most water-stressed rivers or aquifers around the world, agriculture in the watershed accounts for most of the water that is withdrawn from the river. One-third of the water used on farms drains back to the river as return flow after being used. The remaining two-thirds is taken up by crops, enabling them to grow, or evaporates from farm soils and irrigation ditches. As a result, agriculture accounts for more than half of all water consumptively used from the Colorado River.[6]

Water budgets, such as the one illustrated in figure 2-5 for the Colorado River, can be very useful in identifying strategies for rebalancing a water budget. In chapter 3, some of the most commonly used strategies for increasing water supplies or reducing consumptive losses will be described. In water-scarce settings, it always makes sense to look for ways to reduce consumptive losses in the largest consumption categories, as part of the overall strategy for alleviating scarcity. As suggested by figure 2-5, reducing consumptive use in agriculture deserves serious consideration in the Colorado River watershed, given that this is the largest category of water depletion. Additionally, one-fourth of the river is diverted out of the watershed (see "exports" in fig. 2-5) to supply water needs in distant cities, so any reduction in water use in those cities can also reduce the need for these exports.

The High Plains Aquifer: Busting the Groundwater Budget

The High Plains Aquifer is one of the world's largest aquifers, spanning an area of 450,000 square kilometers (174,000 square miles) across eight states in the central United States (fig. 2-6).[7] The High Plains is a massive reservoir of sand and silt that was deposited millions of years ago as the Rocky Mountains eroded and sediments were washed or blown into ancient valleys, filling them to depths of more than 400 meters in some places. That earthen reservoir subsequently filled with water, with a surge of recharge during the last ice age, creating an underground store of water that is more than 300 meters deep in some places.

Early settlers struggled mightily to grow crops in the semiarid Great

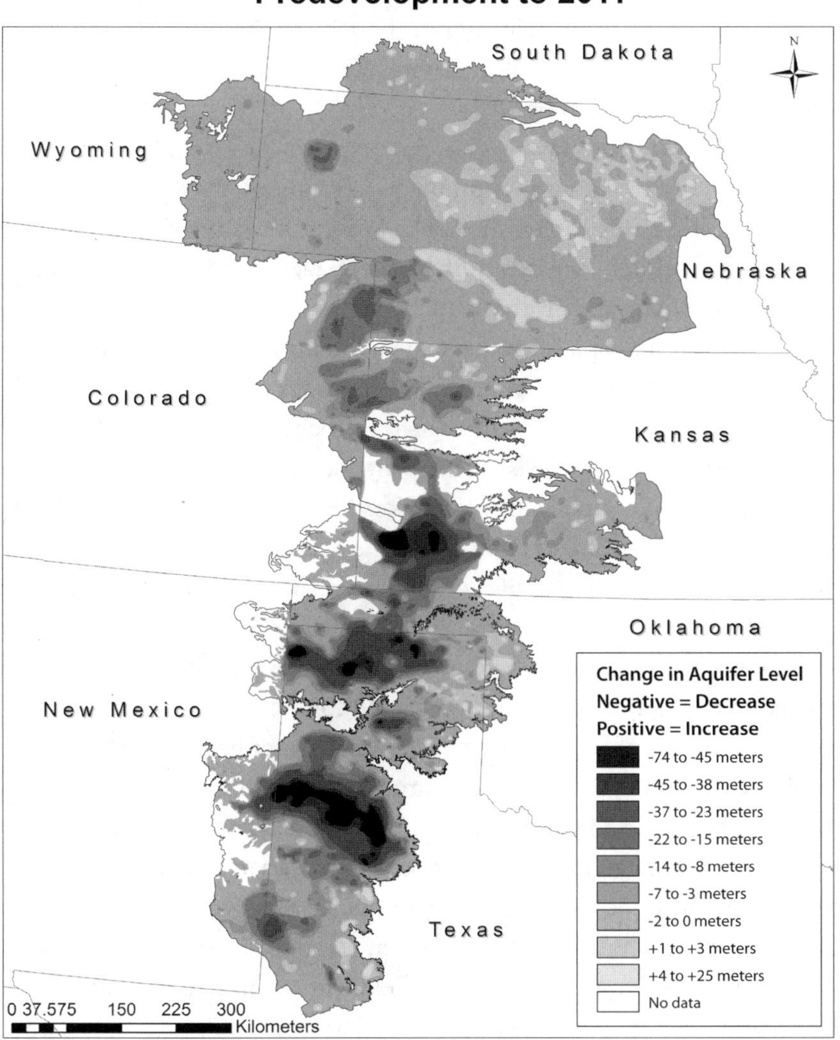

Figure 2-6. *Water level changes in the High Plains Aquifer, predevelopment to 2011. (Based on data from the U.S. Geological Survey.)*

Plains, where precipitation ranges from an annual average of 300 millimeters (12 inches) in the west to 800 millimeters (33 inches) in the eastern areas overlying the aquifer. But as in many other parts of the world, the availability of cheap, efficient electric pumps after World War II made it possible to access the High Plains' vast stock of underground water.

Agricultural irrigation using groundwater spread rapidly to nearly 60,000 square kilometers (23,000 square miles) over the High Plains Aquifer—comprising a fifth of all cropland in the United States—creating one of the most productive agricultural areas in the world for growing corn, wheat, soybeans, and livestock.

But the High Plains farmers are now experiencing a very serious water bankruptcy because they are pumping water out of the aquifer ten times faster than it is being replenished. The amount of water deposited in the High Plains Aquifer is greatly limited by the fact that most of the meager precipitation that falls on the land evaporates very quickly. Only a tiny fraction of the precipitation percolates downward into the aquifer. The resultant annual recharge of the aquifer ranges from almost nothing in the south to only 15 millimeters (half an inch) to the north. With such small water deposits, it is easy to see why the aquifer is being overdrafted.

Unlike the situation with rivers like the Colorado, very little of the water withdrawn from aquifers is returned to the aquifers after use. The water pumped from an aquifer might run off farm fields or down urban drains, but it usually ends up in a river instead of being returned to the aquifer after use. In a water budget for the aquifer, any water that is not returned to the aquifer would be considered to be consumptively used with respect to the aquifer. Consumptive losses of water have caused the water level in the High Plains Aquifer to drop by more than 1.5 meters (5 feet) per year overall, and by more than 70 meters (230 feet) in some places.

In total, the High Plains Aquifer has lost less than 10 percent of its water volume since the 1950s. That might not seem like much of a problem, but here's the catch: As the water level in an aquifer drops, the cost of the electricity needed to lift it onto a farm field with a pump rises sharply. Many farmers in the region can no longer afford to irrigate. For them, the water from the High Plains Aquifer is as far out of reach as water from the Congo.

Because 95 percent of the water pumped from the High Plains Aquifer goes to irrigated agriculture, this is the obvious place to look for solutions to the overdraft of the aquifer. There is no way to arrest the decline of the water level without substantially reducing agricultural irrigation. Many farmers in the area have begun using more water-efficient technologies and practices that have reduced their withdrawals by 20 percent or more, but given the reality that current pumping levels still greatly exceed the recharge rate of the aquifer, the question is not if the aquifer will be depleted, but at what rate.

Be Careful with Averages

The water accounting presented above, for the Colorado River and the High Plains Aquifer, is based on long-term averages; this method evens out the deposits, withdrawals, and consumptive uses that in reality might vary greatly from month to month or year to year. The deposits into a watershed or aquifer account are not fixed like regular deposits from salary paychecks. Instead, they are more like the irregular income that an independent consultant earns, which can vary considerably over time, depending on workload.

Returning to the bank account analogy, when your income (or precipitation) is high, you might be able to save a little for drier times. That is why huge water storage reservoirs such as Lake Mead were built on the Colorado River—that reservoir alone can store the equivalent of 2 full years of the Colorado's flow. But when you hit a bad stretch with lesser income or precipitation, if you do not have savings in reserve, you could get into a lot of trouble.

Most of the water users experiencing water shortages do just fine during normal or wet years. But during dry years or prolonged droughts, they get into serious trouble because, just as with your family bank account, it can be very difficult to cut back water consumption sufficiently and quickly enough to avoid going bankrupt. In some instances, communities that cannot get any more out of their rivers have switched to pumping from aquifers, or importing their water from other watersheds. But in too many instances, simply switching to a new water source without constraining overall consumptive use only spreads water bankruptcy to another place. For example, after Los Angeles dried up its namesake river and depleted its local aquifers, it began drying up Mono Lake and the Sacramento River in Northern California as well. After farmers and cities in northern China depleted both the Yellow River and the aquifer beneath the North China Plain, water had to be imported from the Yangtze River basin hundreds of kilometers to the south at the cost of more than $60 billion.

Planning for a Secure Water Future

It makes great sense to develop a long-term plan to ensure your financial security. In doing so, you will need to anticipate both short-term and long-term changes in your income, and how much you expect to spend on an average basis as well as for occasional bigger purchases such as buying a car. You may also want to keep some money in savings to avert shortfalls when unexpected changes arise in your income or expenses.

These sound financial principles are equally pertinent in water planning. Of particular importance will be projecting changing demands on the available water sources, such as those from population growth, expansion of agricultural areas, or increases in industrial or energy development. It is also of critical importance to gain an understanding of the likely variability or trends in water availability, particularly given climate change forecasts of substantial changes in precipitation or evaporation in many regions. For instance, climate scientists are now projecting that water flows in the Colorado River could decline by 5 to 20 percent in coming decades, suggesting even-greater challenges in alleviating water scarcity in that watershed.

Rain and snow volumes can also vary considerably from season to season and year to year. Water plans must explicitly address strategies for balancing the water budget with more or less water deposited in the account.

Squandering Our Inheritance

If your water source is a river and the rate of consumptive use has reached the rate at which the river is being replenished by precipitation and runoff in the watershed, the river is dried up and there is no more water to be used. In that sense, there is an ultimate physical limit to the volume of potential consumptive use from year to year. More water will keep coming from precipitation, but you cannot consume faster than you receive for very long without bottoming out the river.

Things are quite different when an aquifer or a large lake is your water source. The rate of consumptive use from an aquifer or lake can exceed the rate of replenishment considerably, for a pretty long while, before water users will experience problems. This is because aquifers and lakes hold a lot of water in storage. It is like having a savings account at the bank that you can plunder after you have spent all the money in your checking account.

There are no signs more telling of the widespread and chronic mismanagement of our planet's water accounts than the ongoing depletion of our aquifers and lakes. Much of that water accumulated over thousands of years, like an inheritance from many past generations, and now we are depleting it in a matter of decades. It is like burning down your house to stay warm for a little while longer.

Global water modelers have estimated that we are now depleting our planet's aquifers at the rate of more than 200 cubic kilometers per year. That is considerably more than the entire volume of Lake Tahoe in the United States or the Dead Sea in the Middle East, about the same volume as all of Lake Turkana in Kenya or Lago Argentino in Argentina, or about half the volume of America's Lake Erie. *We are depleting that much each year.*

Speaking of lakes, we are heavily depleting many of them as well. In less than 50 years, virtually the entire Aral Sea—once the world's twelfth-largest lake—was dried up as the former Soviet Union tried to grow rice, melons, and cotton in the desert surrounding the lake.

Just as with the gradual depletion of a financial inheritance or savings account, we can spend more than we deposit, consume more water than is replenished, for a little while. But eventually, the water level in an overpumped aquifer drops so low that we cannot afford the electricity to pump it from such great depths, or the shoreline in an overused lake recedes so far that boat docks and irrigation canals are left high and dry.

In the Hindi language, the words used to describe money are differentiated according to the original source of the funds. *Aap kamai* is used to describe one's own income, but an inheritance is referred to as *Baap kamai*. There is a common Hindi saying that one must limit one's spending within one's own income and not eat into the inherited assets, otherwise bankruptcy will soon appear.

That philosophy is aptly applied to water. The next chapter describes some of the most commonly used approaches to balancing a water budget, by either increasing water supplies or reducing consumptive uses. There is no one perfect mix of supply and demand strategies; instead, the available approaches must be custom tailored to each water account.

CHAPTER THREE

Options for Resolving Water Bankruptcy

As the year 2012 drew to a close, the U.S. Bureau of Reclamation published the results of a comprehensive study of the Colorado River watershed in the western United States. The final report of the Colorado River Basin Water Supply and Demand Study summarized more than 150 different ideas for balancing the water budget of the Colorado River. One of those ideas grabbed headlines across the country: a scheme to build a water pipeline 1,000 kilometers (620 miles) in length from the Missouri River to Denver.

Pipeline boosters argued that the water-import project could replenish overdrawn watersheds and aquifers all along the pipeline's path and relieve pressure on the Colorado River by providing an alternate supply for cities like Denver that depend heavily on transmountain imports of water from the Colorado River's watershed.

For many, the proposal conjured memories of a grand plan called the North American Water and Power Alliance, or NAWAPA. That plan—conceived in the 1960s by the Ralph M. Parsons Corporation, a giant engineering firm based in California—envisioned diverting water from rivers in Alaska and then moving the water south through Canada to eventually rewater the parched American Southwest. The proposed water transport and storage system would have spanned more than 3,000 kilometers (2,000 miles), requiring hundreds of separate construction projects, with a price tag of $100 billion. According to its proponents, the project would have doubled the total amount of freshwater available to the lower 48 states, forever solving the water shortage problems of the western United States.

NAWAPA was never built, but its boldness clearly inspired Chinese engineers, many of whom studied in U.S. engineering colleges. As the aquifers and rivers of the North China Plain—China's breadbasket—were increasingly depleted during the latter half of the twentieth century, Chinese engineers began implementing Chairman Mao Tse-tung's decades-old vision of moving water from China's water-rich southern region to the much-drier northern region where most of the country's agriculture is located. The water transfer is to be accomplished with a trio of long-distance, high-volume canals. The Chinese are currently investing $62 billion to build this system—known as the South–North Water Transfer Project—to move water over thousands of kilometers from the Yangtze River to water-stressed cities and farms to the north. As the *New York Times* reported in 2011, "It would be like channeling water from the Mississippi River to meet the drinking needs of Boston, New York and Washington."[1]

Is moving water across great distances the answer to water shortages?

The Saudi Arabians have taken a different course. That country has been the world's heaviest investor in desalination technology, which is used to remove salt from seawater, turning it into freshwater. After heavily depleting their groundwater aquifers during the latter half of the twentieth century, primarily for the purpose of growing wheat, the Saudis had built thirty desalination plants by 2009. The plants provide half of the freshwater supply in Saudi Arabia and represent half of the world's total desalination capacity.

Is removing salt from seawater the answer to freshwater scarcity?

The "right" solution to any community's water shortage problems will depend on different factors: cost, environmental impacts, energy requirements, land ownership, and other considerations. Disagreements over proposed solutions to water scarcity are quite commonplace. From my observations, I attribute most of these disagreements to two primary factors: widespread water illiteracy (i.e., debates that are too often based on ignorance of what is feasible or sufficient), and inadequate or improper decision-making processes for choosing among available options, a problem I will begin to discuss in the next chapter. In this chapter, I will describe a "toolbox" of options available to communities that want to increase their water supply or reduce their water use and consumptive losses, and I will illustrate how some communities have applied those tools.

Increase Supply or Reduce Demand?

Returning to our bank account analogy, there are two basic ways to balance a budget: increase deposits or reduce expenditures. This is the crux of any plan to resolve a water shortage: should you find new sources of water supply or reduce your use?

The per-unit cost of either supplying or saving water (e.g., dollars per cubic meter) is almost always the most prominent influence in plans for addressing water shortages. However, other important factors can, and should, complicate the decision-making process. There are almost always environmental impacts or benefits that must also be weighed in any water plan, and in many cases, there are social consequences in addition to financial concerns that must be taken into account. Unfortunately, special interests and widespread corruption among both governments and private contractors too often create an uneven playing field, a problem that I will address in the next chapter.

The Water Toolbox

There are six general options for balancing a community's water budget. Some involve tapping into new sources of water, some involve storing water to alleviate seasonal or temporary shortages, and some involve reducing the amount of water that must be withdrawn or consumptively used. I will describe each of the six options here, in the general order of highest to lowest cost: desalination, water reuse, water importation, water storage, watershed management, and water conservation.[2]

Desalination

This is a technological process that involves removing salts and other minerals from either seawater or brackish (salty) groundwater. The traditional way of doing this has been to use a distillation process that involves boiling the salty water to separate (evaporate) the freshwater from the salts, but recently the use of reverse osmosis has become a preferred method because it is less expensive. In reverse osmosis, the salty water is pushed through a semipermeable membrane that allows water molecules to pass through but not salts or other minerals. Typically, about half of the salt water processed in desalination ends up becoming freshwater, leaving behind a heavily concentrated waste known as brine.

Desalination can be a feasible and attractive option for creating additional water supply along coastlines or in places where brackish aquifers can be readily tapped. The great benefit of desalination is that it protects natural freshwater sources—rivers, lakes, and aquifers—from being depleted further. As coastal populations grow, a shift toward increased use of desalination could go a long way toward alleviating pressure on other freshwater sources.

The biggest disadvantage is cost; it remains the most expensive means for obtaining water supplies. The high cost of desalination is due to the great amount of electricity it requires. However, desalination of brackish water is usually considerably less expensive than desalting seawater, because of the typically lower salt content of brackish water.

Generating electricity to power desalination plants can also produce carbon emissions that are causing climate changes. Until recently, Saudi Arabia used only oil and natural gas to generate the electricity needed to power its desalination plants; the country has been using 1.5 million barrels of oil each day to generate energy for desalination.[3] However, the Saudis are now building the world's largest solar desalination plant, and they have expressed intent to eventually convert all of their desalination facilities to solar power.[4] Similarly, because of climate change concerns, a new desalination plant in Adelaide, Australia, was recently built using 100 percent renewable sources of energy.

Properly disposing of the brine leftovers in an environmentally safe and cost-effective manner can also be a serious challenge. In coastal environments, the brine is usually discharged some distance offshore, where ocean water can quickly dilute it, but it can damage sea life if not dispersed properly. Disposing of the brine after desalting inland brackish water usually involves injecting it into deep, already saline aquifers, which can be very expensive and can only be done where such saline aquifers exist.

There are now nearly 16,000 desalination plants in the world, with the largest found in Saudi Arabia, the United Arab Emirates, and Israel. The use of desalination is growing rapidly, but all of the plants in the world today still supply less than 1 percent of all freshwater withdrawals. Although desalination can be an important strategy for alleviating local water scarcity—such as in Saudi Arabia where aquifers have been heavily depleted and no perennial rivers exist—its potential to resolve water scarcity more broadly will be quite limited until the energy demands of the technology are reduced substantially and brine disposal challenges are resolved.

Water Reuse

Also known as water recycling, water reuse involves purifying water after it is used in homes, businesses, or industries and then putting the water to another use. In the reuse process, impurities are removed from the wastewater to a level appropriate for its intended reuse. Most commonly, the reused water is applied to farms, golf courses, and other landscape areas and is sometimes used in power plant–cooling facilities or other industrial processes. Israel is the global leader in this process, reusing about 80 percent of all water it withdraws from freshwater sources. The next-largest reuser is Spain at just 17 percent, and all other nations are much lower.[5]

There are few places where reused water becomes drinking water again, primarily due to psychological aversion, rather than legitimate water quality concerns (this has been referred to as the "yuck factor," a common response to the thought of drinking wastewater, no matter how clean). Given that 6 percent of the water flowing in U.S. rivers has already been used, treated, and then discharged as return flow from upstream cities, many Americans are in fact already drinking someone else's processed wastewater.[6] With proper treatment, most wastewater could be safely reused for drinking purposes. For example, Singapore delivers recycled water in bottles, or pipes it into homes or businesses, under the brand name of NEWater. And astronauts traveling in space recycle their urine and other wastewater and drink it aboard their spaceships while in orbit, so just about anything is possible!

The cost of reusing water is usually high for the same reason that desalination is expensive: The process of removing undesirable substances from wastewater can require a great deal of energy. However, reuse technologies are increasingly being customized to produce treated water at a quality level appropriate for the intended end use, which can be much less expensive than always treating the water as though it were going to be used for drinking. Important to note here is the fact only 1 to 3 percent of all the water used in a city is actually used for drinking, so considerable opportunity exists to recycle water at lesser cost than needed for drinking water.

In many industrialized countries, reuse is an appealing option when it is difficult or costly to further exploit other freshwater sources. It is most popular in urban areas that are growing rapidly, where recycled water can be applied to landscape watering or industrial purposes. In developing regions with fast-growing populations, reuse is also providing an important source of water for expanding agricultural irrigation. Reusing water can also help alleviate water quality problems if the wastewater would have otherwise been returned to a river or lake with inadequate or no treatment.

It is very important to understand, however, that recycled water is not truly a *new source* of water. Instead, it is simply a way of making more use of the same water. From the perspective of water scarcity, recycled water must always be viewed in the context of the local water budget (fig. 3-1).

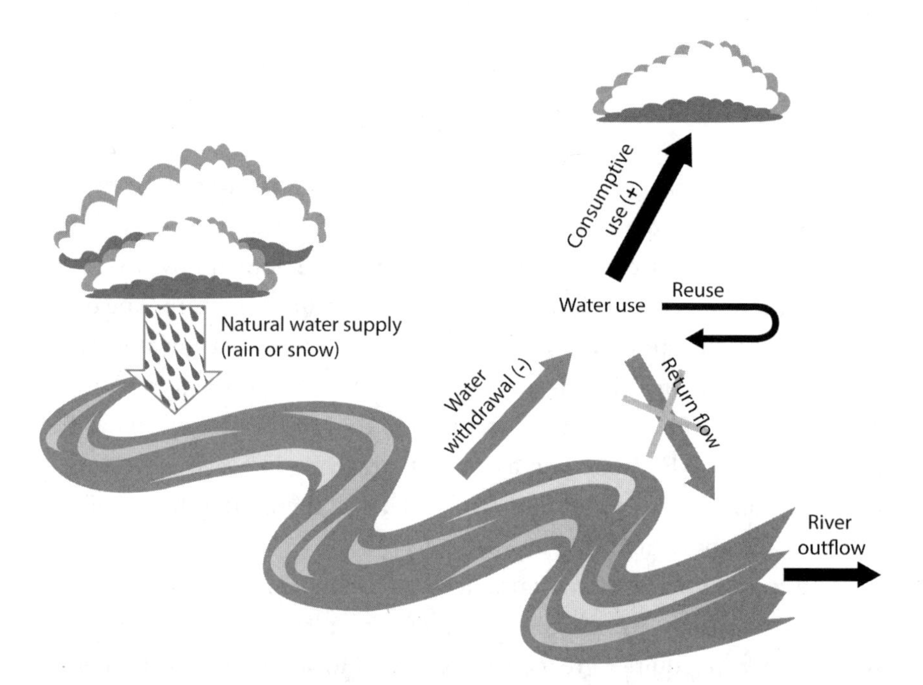

Figure 3-1. *It is very important to view water reuse (also known as water recycling) in the context of a water budget. If water is reused, the volume of water that must be withdrawn from rivers, lakes, and aquifers can be minimized. However, water reuse intercepts water that may have otherwise been returned to natural sources. If water is no longer returned to the water source after use, the result can be increased net consumptive use—and increased depletion—of the original water source.*

Water reuse intercepts return flows that would otherwise partially replace the volume of water being withdrawn from freshwater sources, and it almost always results in further net depletion of a water source. Instead of returning wastewater to a river or lake after it is used in homes and factories, reuse redirects water to another use such as landscape watering that may then consume the recycled water in its entirety. This can result in net *increased* consumptive loss, exacerbating water shortages. On the other hand, water reuse can be important in helping to reduce groundwater consumption, because that water is almost never returned to the aquifer anyway, and reusing water may lessen the volume of water that needs to be withdrawn from aquifers. In most instances, all groundwater withdrawals result in a consumptive loss, so any reduction in pumping can help reduce aquifer overdraft. The same would be true of a city or industry using water from a river or lake but sending its return flows to a different water body or to the ocean.

Water reuse makes a lot of sense if there are times, such as in droughts, when a city or industry simply cannot withdraw enough water from a river because it is too low, or because the city does not possess the rights to use additional water. By reusing water, a city will not need to withdraw as much water from a river, lake, or aquifer, making the city less vulnerable to low water levels or competition for limited supplies.

Water Importation

For thousands of years, as the water needs of cities and farms have exceeded what their local water sources can supply, they have secured supplemental water by reaching into other water sources. Hundreds of years before the Common Era (BCE),[7] water use in Rome had heavily depleted the flow of the Tiber River, which runs through the heart of the city. Beginning in 312 BCE, the Romans began building aqueducts to bring additional water to the city. When the ancient city's population reached 1 million, eleven different aqueducts were delivering nearly 1,000 liters (260 gallons) per person each day, more than what is available to urban dwellers in most cities today. Some of that water came from nearly 100 kilometers (60 miles) away.

Unlike the Roman aqueducts that were designed to move water using only gravity, most water importation projects today require huge amounts of energy to push water uphill along portions of their routes, and this makes the water very expensive. In the United States, the California State Water Project—which moves water from Northern to Southern California—and the Central Arizona Project—which moves water from the Colorado River to Phoenix and Tucson—are the biggest electricity hogs in those two states.

Another big concern about water importation—also commonly known as interbasin transfer—is the fact that it can spread water scarcity or intensify it in other watersheds or aquifers. A case in point: Early in the twentieth century, Los Angeles had fully consumed its namesake river. The city then built a 500-kilometer (300-mile) pipeline northward into the Owens River and then extended it into the watershed of Mono Lake, depleting those water sources and damaging their ecological health. After exhausting those water supplies, the city extended its long straw into the Colorado River far to the east, and into the Central Valley rivers farther up north, contributing to water scarcity in those watersheds.

Water can also be imported virtually, through the trade of consumer goods that require water in their production. The term *virtual water* was originally coined by Professor Tony Allan of King's College in London to recognize that each piece of fruit, the barley and hops that go into a glass of beer, and the cotton in a shirt all require water to grow or be manufactured. Arjen Hoekstra, once Allan's student, began quantifying those volumes of water as a product's "water footprint"—for example, 185 liters (50 gallons) to grow an apple; 300 liters (80 gallons) of water for a liter (quart) of beer; 2,500 liters (660 gallons) for a cotton shirt.[8] Allan pointed out that when we ship these goods from one place to another, we are in effect transporting virtual water. By purchasing these goods from other places, we avoid using local water to make them. However, as with the transport of actual water, consumers should be mindful that use of water to grow or produce these goods may be depleting far-distant watersheds.

Taking water from other watersheds or aquifers diminishes the potential for communities in those other watersheds to fully utilize those water sources for their own purposes, and it can greatly damage the health of freshwater ecosystems in those places. For this reason, many governments have instituted "basin of origin" protections that limit or prohibit such transfers of water from one watershed or aquifer to another. For example, when the Ontario government in Canada issued a permit in 1998 to a company seeking to ship 600,000 cubic meters (160 million gallons) of Lake Superior water each year to Asia, the resulting public outcry was so strong that it catalyzed an international agreement among eight U.S. states and two Canadian provinces sharing the Great Lakes. The Great Lakes Charter Annex, passed in 2001, makes large-volume exports of water from the Great Lakes highly unlikely and intense scrutiny a certainty.

While water importation has some serious economic, environmental, and social drawbacks, it has been used to alleviate local water shortages for a very long time, and it continues to be used today. By some estimates, more than 350 large-volume water transfer schemes have been constructed in the past 60 years. Metropolises such as Los Angeles, Denver, New York, Mumbai, Karachi, Tel Aviv, and Cape Town would have likely remained small cities if they had not built pipelines and canals to import water from other watersheds. The continued growth of Tianjin and Beijing, and China's vast agricultural irrigation in the North China Plain, will be supplied for the foreseeable future by transferring water from the Yangtze River through the South–North Water Transfer Project. Similarly, in the most extensive water transfer scheme ever proposed, India has designed a "river-linking project" that would connect 37 different rivers using 9,000 kilometers (5,600 miles) of canals, at an estimated cost of $140 billion, as the country's solution to its water shortages.

Water Storage

There are many places in the world where sufficient water is available on an average yearly basis, but shortages arise during certain times of the year. Imagine a teacher who receives a salary from her school only 9 months of the year—she gets paid nothing during the summer break. She might make enough in 9 months to meet her needs over the whole year, but she will need to save some money during the school year so that she will have some money to use during her break.

The same challenge exists in managing water in places like the Santiago River in Mexico (fig. 3-2). Water is plentiful in this watershed during winter and spring, but during the summer growing season, water consumption in irrigated agriculture is much greater than what the river can naturally provide. For that reason, dams have been built in the watershed

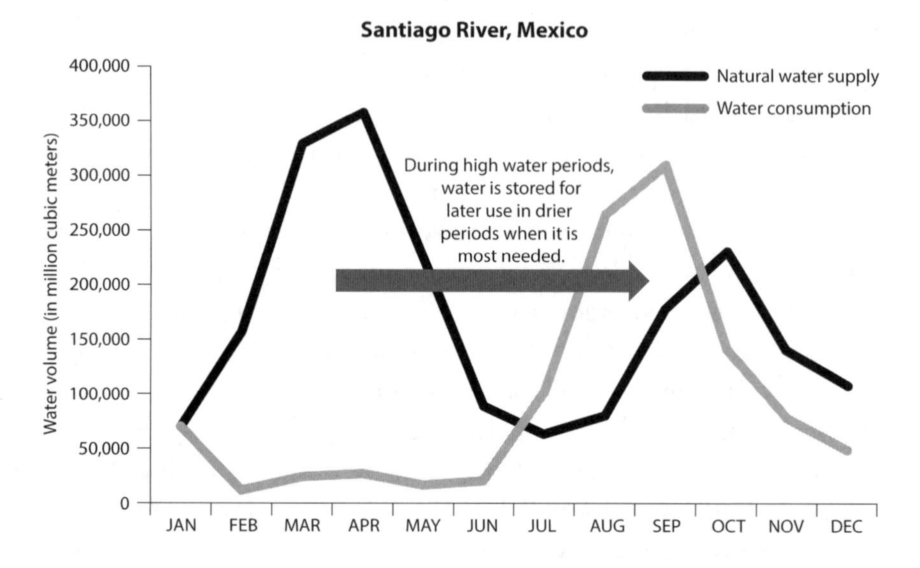

Figure 3-2. *In the Santiago River of Mexico, water needs are greatest during the summer months, to support irrigated agriculture. However, most of the annual water supply in this watershed comes from winter and spring rains. To resolve this disparity in the timing between water supply and use, large reservoirs have been built to capture and store water during the rainy season so that it can be used in the summer growing season. (Based on data from UNESCO-IHE Institute for Water Education.)*

to capture and store water from high flows during the winter and spring and hold it for later use in the summer.

There are many disadvantages to building dams to store water, however. One is their cost. In the range of options discussed here, dam building falls in the midrange of cost-effectiveness. Dams and their associated reservoirs can have huge environmental and social impacts. They are a leading cause of the decline of fish and other river species globally because they block their movements and change the flow of water, nutrients, and sediments through the river ecosystem. They have disrupted the lives of hundreds of millions of people who lost their way of life or even their homes when dams were built.[9] Dam-created reservoirs can also lose a lot of water to evaporation, particularly in arid areas. Fifteen percent of all consumptive use of water in the Colorado River basin of the western United States is due to evaporation from reservoirs, for example.

Because of evaporation losses from dam-created reservoirs, storing water underground for later use is smart, and gaining in popularity. Sometimes referred to as "conjunctive use" or "aquifer storage and recovery," this strategy usually involves injecting water extracted from a river or lake during high-water periods into an aquifer, and then later pumping the water back out of the aquifer for use during periods when less water is available in the river or lake. Conjunctive use does not necessarily require the use of dams to temporarily slow or store the water before injecting it into an aquifer, but this is the usual case.

The practice of "rainwater harvesting" has long been practiced in India, as well as in other parts of Asia and in Africa. This typically involves building a small rock-and-soil dam in a stream channel to capture water runoff during rainstorms. Sometimes the water is directly withdrawn from the small dam-created reservoir, but more common is to allow the captured water to percolate underground, where it collects in a shallow aquifer and can be accessed from hand-dug or deeper wells. Similarly, "stormwater capture" is being used in many cities where concentrated runoff from urbanized areas is captured in small reservoirs, where it can percolate into an aquifer and later be retrieved.

When contemplating the use of a storage reservoir to alleviate water shortages, one should always carefully evaluate the impacts on the watershed or aquifer water budget (along with the other possible impacts discussed above). If the sole purpose of the reservoir is to shift the timing of water availability—such as in the Santiago River illustrated in figure 3-2—then the impacts on the water budget will be beneficial on a seasonal basis. However, when a reservoir is used for the purpose of capturing more water to enable an increase in overall consumptive use, then its impact on the water budget will most certainly be negative.

Steve Leitman, a veteran of a long-running debate over water use in the Apalachicola-Chattahoochee-Flint watershed in the southeastern United States, draws on the bank account analogy when discussing reservoirs. "If you're spending more than you're earning, simply opening up new bank accounts isn't going to make things any better." In other words, reservoirs do not really create any new water. Instead, they simply capture the water supplied by rain or snow and hold on to it until needed.

Watershed Management

The vegetation and soils found in a watershed influence the way that water moves through it or into aquifers. Some types of plants will use more water than others, and plant leaves of different sizes and shapes will intercept precipitation in differing amounts, resulting in different rates of evaporation from the leaves. These biological and physical conditions and processes in the watershed can be manipulated in various ways to influence local hydrologic cycles, thereby changing the yield of water from a watershed, the rate of aquifer recharge, or the quality of the water that runs off the landscape.

There are many watershed management strategies available for influencing the flow of water and its quality, and many are highly cost-effective in making more water available. Removing deep-rooted, water-consumptive shrubs or trees and replacing them with grass cover can free up water. Many communities are also restoring the natural functioning of wetlands and floodplains—for example, by removing levees and allowing rivers to spill onto low-lying lands during floods—in an effort to slow floodwaters and induce more recharge into aquifers.

In South Africa, more than 9,000 different types of exotic plant species have been introduced from other countries, such as thirsty eucalyptus trees from Australia. These introduced plants have changed the landscape and biodiversity of South Africa, and they have also changed the water budget of the country's watersheds. The Department of Water Affairs in South Africa estimates that these vegetation changes have resulted in an average of 7 percent more consumptive loss of water overall, thereby reducing the amount of water available for other uses. A program called Working for Water has employed more than 20,000 South Africans in removing undesirable vegetation from more than 1 million hectares (2.5 million acres). In some watersheds, these actions have improved river flows and water availability by more than 20 percent.

Water Conservation

If you are going into overdraft month after month at the bank, any reputable financial advisor will tell you that it is time to start spending less money. There is no smarter and cheaper way to alleviate water shortages than to apply the same logic. Because of the cost-effectiveness and environmental benefits of water conservation (i.e., less water is taken from rivers, lakes, and aquifers), *water conservation should always be the first place to invest, and its potential should be maximized before the other tools discussed above are deployed*. But you have to be smart about what type of water conservation you apply.

By "smart" I mean that you need to pay attention to where the largest volumes of water withdrawal or consumptive use are occurring in the water budgets of the watersheds or aquifers you depend upon. There are two ways to apply water conservation effectively. First, invest in measures that reduce consumptive water losses in your watershed or aquifer, thereby making more water available for people and nature. If your water source is a river, you should assess how much water is being consumptively used in the watershed upstream of your location, such as in irrigated agriculture. Similarly, if your water source is a lake or an aquifer, you should assess how much of the water being used is not returned to the lake or aquifer after use. Where are the greatest volumes of water being lost? Can those consumptive losses be reduced through water conservation measures so that more water is available at your location? In almost all cases, the smartest investments in water conservation will be directed at those types of water use that are resulting in the greatest consumptive losses.

Second, do what you can to reduce your need to withdraw water from freshwater sources. Are there times of the year, or certain years, during which there is insufficient water available for your withdrawal needs? In the vicinity of your water use, who is withdrawing and using the greatest volumes of water? Can you or other water users reduce water withdrawals by implementing more efficient ways to use water, so that the pressure on the stressed water source can be reduced?

Leaky pipes should be one of the first places for cities to look for opportunities to reduce water withdrawals. The U.S. Environmental Protection Agency estimates that about 17 percent of all water distributed to homes and businesses in the United States is lost due to leaky pipes. Boston loses 30 percent of its water, and London loses almost 50 percent. Some of the biggest water users in your home are the washing machine, toilet, and shower or bath, so those are places to be particularly careful with water. Look for the most water-efficient brands, and use them for the shortest time possible.

The greatest overall gains and cost-effectiveness in water conservation will be realized in reducing consumptive losses in irrigation, both in the city and on the farm. Large irrigated landscapes can be found in many of the cities located in semiarid or arid climates where evaporation losses are extreme. Water is poured onto parks, golf courses, and vast residential and commercial lawn areas, and unlike the water that is used indoors, virtually all of that outdoor landscape water is fully consumed—almost none of it returns to the water source after use. In many cities, more than half of all water use goes to outdoor landscape watering, which largely explains why the per capita water use rates for people living in the American Southwest are among the highest in the world. For conservation models, these cities can look to San Antonio, Texas, which had cut its outdoor use by 30 percent between 1980 and 1995, or the Irvine Ranch Water District in California, which cut its outdoor use by 46 percent from 1992 to 2004.[10]

Australian cities are among the world's best in water conservation, using half as much as cities in the western United States on average. Outdoor watering restrictions and planting of drought-tolerant plants are the primary reasons for this difference. Even when water is not in short supply, many Australian cities limit outdoor water use, and if residents notice neighbors wasting water, they can call a hotline to report water wasters, who must pay steep fines.

By simply planting less-water-intensive or drought-tolerant vegetation in place of water-guzzling grass lawns, substantial water savings can be realized. During the 1990s, more than 60 percent of all residential water use in Las Vegas, Nevada, went to landscape areas. Beginning in 1999, the city began a "cash for grass" program that paid homeowners nearly $17 per square meter ($1.50/square foot) to remove their grass lawns and replace them with desert vegetation that does not require watering. One of the homeowners who took the cash—but probably didn't need it—converted the 6.5 hectares (16 acres) surrounding his 29-bedroom home into desert landscaping and now saves 17 million liters (4.5 million gallons) every year! By substantially reducing outdoor watering, Las Vegas was able to lower its water use by nearly a third in the last decade.

Using drip irrigation—in which water is delivered directly to plant roots through plastic tubes—instead of sprinkler irrigation can greatly reduce outdoor water use in cities. With sprinkler irrigation, much of the applied water evaporates before reaching the ground, and some of the water may be sprayed onto streets or sidewalks or into other areas that should not be watered. Interestingly, one research study found that even hand watering used nearly 70 percent less water than conventional sprinkler systems simply because water was being applied only where needed, and only for as long as needed.[11]

While important reductions in consumptive use can be gained in cities, they generally pale in comparison to the opportunities to alleviate water scarcity through agricultural water conservation, simply because the volume of water used on farms is so much greater. As highlighted previously, in most water-scarce watersheds around the world, more than 90 percent of all consumptive use goes to agricultural irrigation. Water scarcity could be substantially resolved in most of those watersheds with a 15 to 20 percent reduction in irrigation consumption.

In a paper we published in the journal *Water Policy* in 2013,[12] my colleagues and I pointed out that the most promising and cheapest means for cities to reduce their water scarcity risk is to help farmers to reduce their consumptive losses, by taking such actions as lining irrigation ditches with concrete, implementing more efficient irrigation technologies, shifting to less-water-intensive crops, and using other water-saving measures. For example, San Diego, California, negotiated a water conservation and transfer agreement with the Imperial Irrigation District in Southern California, in which the city pays farmers for implementing water conservation measures. The Colorado River water no longer used on Imperial Irrigation District farms—which will increase in volume between 2003 and 2021 from 12 to 247 million cubic meters (3.2 to 65.3 billion gallons)—will be transferred to the city. These agricultural conservation measures will provide 37 percent of San Diego's water supply by 2020.

One aspect of water conservation that deservedly gets a lot of attention these days is water pricing. Unsurprisingly, most people will use less water if they have to pay more for it. For this reason, many cities have instituted water pricing schemes that charge urban residents more as they use greater volumes of water. However, great care must be taken to ensure that such pricing does not make water unaffordable for the poorest members of our society.

Applying the Toolbox in the Colorado River Basin

The Colorado River Basin Water Supply and Demand Study, discussed at the beginning of this chapter, illustrates some of the options for alleviating water shortages. To prepare the study, the U.S. Bureau of Reclamation invited regional water agencies, water experts, and other stakeholders to submit their ideas for either increasing water supply or reducing consumptive uses of water. More than 150 different ideas were submitted, including 55 options for increasing water supply, 42 options for reducing consumptive losses, 22 options related to modifying the operations of dams and other infrastructure, and 41 other options for improving management. Some of the options that received highest priority are summarized in table 3-1.

It is not difficult to identify which of the tools listed in the table should be prioritized for their contribution in reducing water shortages in the Colorado River watershed. Some of the options are cost-effective and can save or produce a lot of water. The five most cost-effective options, listed in order of expense, starting with the least expensive, include weather modification, agricultural water conservation, forest management, urban water conservation, and desalination.

Three of the tools in table 3-1 could be used to very positive effect. Agricultural water conservation is a clear winner. Its cost-effectiveness, lack of adverse environmental impacts, potential benefits to water quality by reducing polluted runoff, and ability to substantially reduce overall consumptive use within the watershed make it a highly attractive option. Because both urban water conservation and desalination would be applied primarily in cities that lie outside the watershed, application of those options could reduce the amount of water that must be exported out of the watershed.

The Bureau of Reclamation's report gave very low priority to weather modification—which involves spraying silver iodine crystals or dry ice into the atmosphere to induce precipitation—because of its uncertain returns and implementation risks. It is also worth mentioning that any effort to induce precipitation over the Colorado River's watershed would very likely be "stealing" water from other areas where the moisture would have naturally fallen.

Table 3-1 Options Identified for Increasing Water Supplies or Reducing Consumptive Uses in the Colorado River Watershed of the Western United States

Option Category	Specific Proposal	Cost ($/cubic meter)	Potential Water Volume by 2035*
Desalination	Gulf of California	1.70	247
	Pacific Ocean in California	1.50–1.70	247
	Pacific Ocean in Mexico	1.22	69
	Salton Sea drain water	0.81	247
	Groundwater in Southern California	0.61	25
	Groundwater near Yuma, Arizona	0.49	123
Water Reuse	Municipal wastewater	1.22–1.46	247
	Gray water (household)	3.40	219
	Industrial wastewater	1.62	49
	Water produced from coal-bed methane-gas extraction	1.62	123
Water Storage	New water storage reservoirs	1.82	25
	Rainwater harvesting	2.55	92
Watershed Management	Brush control	6.08	62
	Forest management	0.41	247
	Invasive vegetation (tamarisk) control	0.32	37
	Weather modification	0.02–0.05	863
Water Importation	Import from Missouri River or Mississippi River to Denver	1.38–1.87	0**
	Imports into Green River from Bear, Yellowstone, and Snake Rivers	0.57–1.54	195
	Imports to Southern California via icebergs, water bags, and tankers	2.19–2.76	740
Water Conservation —Urban	Municipal and industrial water conservation	0.41–0.73	740
Water Conservation —Agricultural	Agricultural water conservation	0.12–0.61	1,233
Water Conservation —Energy	Power plant conversion to air cooling (rather than water cooling)	1.62	197

* (million cubic meters / year)

**This is due to the inability to complete the project by 2035; by 2060, an estimated 740,000 million cubic meters per year could be available.

Adapted from U.S. Bureau of Reclamation, *Colorado River Basin Water Supply and Demand Study* (Denver: USBR, 2012).

Mastering the Toolbox

As illustrated by the Colorado River example, investing in agricultural water conservation is usually the least expensive and most environmentally friendly way out of water scarcity in the water-short places in the world. But that is not always where the investments are made. There are very few places in the world where the water toolbox described in this chapter is being applied in a way that might be considered optimal or sustainable.

An optimal solution would require balance among economic, social, and environmental values. Water investments in the toolbox would be prioritized according to their cost-effectiveness and applied in a way that creates jobs, fosters maximum economic output per unit of water consumed, and avoids impacts on the poor and freshwater ecosystems. The full array of social values associated with water use would be given their due consideration, and water-sharing decisions throughout the watershed would be equitable, inclusive, and balanced. Ecosystems and species would be protected or restored, benefiting local communities in myriad ways and sustaining the natural heritage of our planet. But that kind of decision making requires an effective water governance system that can foster optimal outcomes—the focus of the next three chapters.

CHAPTER FOUR

Who Is Responsible for Water?

Abu Khalil, a cotton grower forced to abandon his Syrian farm because of a lack of water in 2013, was interviewed by columnist Thomas Friedman for the *New York Times*.[1] Friedman described severe water shortages in Syria as a primary culprit in sparking the tumultuous civil revolt that swept across the country beginning in 2011. "We could accept the drought because it was from Allah," said Abu, "but we could not accept that the government would do nothing." Syrian economist Samir Aita echoed that sentiment, telling Friedman, "The drought did not cause Syria's civil war, but the failure of the government to respond . . . played a huge role in fueling the uprising."

These Syrians are expressing a complaint heard around the world these days, and it is a message that we can no longer ignore: most governments are not performing well in their job of managing water, and the consequences are reverberating through our societies.

In chapter 2 I highlighted the importance of managing our water sources within the limits of water availability, that is, within a water budget. As with a bank account, we cannot consume more than has been deposited without dire consequences. But here is where the analogy with money accounts begins to break down. We each understand that it is our individual responsibility to properly manage our personal bank account, but most of us assume that our government is taking care of our water accounts. After all, in most countries the government decides who gets to use water, and how much—or at least it is supposed to work that way.

There is no question that balancing a water budget and managing water equitably and profitably is very challenging. It requires that someone (oftentimes elected) decides who does and who does not get to use water. That decision-making process is almost always subject to the influence of powerful interest groups, including domineering ministries within government and hard-lobbying corporations demanding more water. Widely divergent opinions and philosophies exist between civil society interests—arguing that water is a public good and access to it a human right—and others insisting that it is an essential input in economic production. Very few people appreciate the enormity of the challenge in balancing these opposing forces. But everyone notices when things do not go well with water.

This book is full of stories of places where governments have been unable or unwilling to take the actions necessary to avert water shortages. Very few governments have put into place adequate controls on water use that will prevent water sources from being exhausted completely or from dropping to such low levels that local economies, livelihoods, and ecosystems are jeopardized. The water shortages in Syria—resulting from a collision of drought and long-running overuse of water for irrigation—had by 2013 disrupted the livelihoods of more than 800,000 farmers. In desperation, many of those farmers flocked to cities like Damascus or Aleppo, where they found little or no relief. Persistent water rationing in these cities provided little water for the wealthy and nothing for the poor, who were often forced to buy their drinking water from privately owned water trucks at exorbitant prices. That awful water situation created a tinderbox for revolt.

Without question, the nations of the world must step up to the task of managing water. The challenge boils down to gaining an understanding of water budgets, selecting the best options for resolving or avoiding water shortages, funding and implementing those options, protecting ecological and human health, and then continuously learning and improving over time. And these activities and decisions must be undertaken in a manner that is perceived to be equitable and fair so that the populace will abide by them. But we should no longer expect nor even ask our governments to do this in isolation from those they govern, for at least two good reasons. The first reason has already been stated: to date, most governments—both national and local—have not been very successful in meeting their water responsibilities. But the better and more important reason for taking a new approach is the fact that history has clearly shown that when local citizens, communities, and businesses are more actively engaged in designing their water future and working through problems, the prospects for success go way up.

Water policy experts around the world are now stressing a need to move away from reliance on centralized, top-down, governmental technocracies in water management and a need to move toward more inclusive and durable *water governance systems*. Progressive approaches to water governance open the circle of water dialogue and decision making to include broader representation from civil society and the private sector. Water governance systems will usually need to include governmental regulation, but they also actively engage citizens and the private sector in designing and implementing water solutions that are equitable, cost-effective, and durable. To better share water, we will need to learn to better share its governance.

To maximize the probabilities for success, any transition to a new system or approach should be founded on a full understanding of the shortcomings and challenges encountered in the old system. I very well remember when, as a teenager, I rebuilt a motorcycle engine in an effort to replace some worn-out parts and install new gears to make the engine more powerful. However, I did not pay sufficient attention to the necessary function and proper placement of some of the other parts when reassembling the engine, and when it was all done, I had a few extra parts lying on my garage floor. The engine never ran well after that. Moving to a new water governance system is a lot like rebuilding an engine. Only by fully understanding the necessary functions that need to be addressed, the roles that need to be played in an integrated system, and what has been going wrong can a new governance system be developed for improved performance and better outcomes.

The Role of Governments in Allocating Water

There is a very important difference between ownership of water and the right to use water. Most countries have taken the position that water should be publicly owned, with the national government acting as the public custodian of water. Many governments presently allow landowners to own the groundwater beneath their land, but recent trends are moving strongly in the direction of public ownership of both surface and groundwater.[2] Accordingly, the right to use water—commonly referred to as "water allocation"—is subject to control and regulation by national governments. It is quite common, however, for national governments to delegate the authority for water allocation to a more local governmental entity, such as a state, province, or irrigation district.

In some countries, the right to use water can take the form of a property right, but it is not a true ownership of the water itself. Instead, it is ownership of the right to use water. It is also important to understand that the right to use water is almost always conditioned with various limitations, as discussed below, and that right can be taken away by the government when it is not used or is used improperly. These are critically important issues to remember when you read a story or see a movie that suggests that a corporation is buying up or taking control of a water supply; in most cases, the government retains the ownership and authority of allocating rights to use the water. This does not mean that powerful individuals or companies cannot exert disproportionate influence over water allocation, but it is important to know that governments usually retain ultimate legal control.

In many countries, water remains sufficiently abundant to meet everyone's needs, and the citizens of the country enjoy open and largely unrestricted access to water sources. However, as the cumulative use of water eventually reaches a point where competition and conflicts begin to develop among water users, or when environmental degradation becomes a concern, a common governmental response is to set up rules for allocating water, such as by issuing rights or "entitlements"[3] to use water.

A fundamental purpose of water allocation rules and regulations is to prevent water bankruptcy—the overuse of water that causes economic, social, and ecological disruption. In 1968, Garrett Hardin, an ecologist, wrote an article in the journal *Science* titled "The Tragedy of the Commons." Hardin explained that a tragedy of the commons results when a shared resource—a forest, a fishery, a water source—is depleted by individuals acting independently and according to each one's self-interest, despite their understanding that depleting the common resource is contrary to the whole group's long-term interests. Hardin's essay has been widely read and cited by political leaders, economists, scientists, conservationists, and many others.

During a trip to central India a few years ago, I witnessed a tragedy of the water commons firsthand. I met a farmer who was installing a large new pump in his groundwater well. I asked him whether he was aware that groundwater levels were dropping rapidly throughout the area surrounding his farm as a result of widespread and excessive groundwater extraction. "Yes, exactly," he replied. "That's why I bought this new pump. I need to get the water before my neighbors do."

Another tragedy of the water commons is presently unfolding in the Gaza Strip, threatening to become a humanitarian crisis with huge international ripple effects. With no streams or rivers to speak of in its territory, Gaza has historically relied almost exclusively on a coastal aquifer, which receives some 50 to 60 million cubic meters (41,000 to 49,000 acre-feet) of refill each year thanks to rainfall and runoff from the Hebron Hills to the east. But serving the water needs of Gaza's rapidly growing population, as well as those of the nearby Israeli farmers, removes 160 million cubic meters (130,000 acre-feet) of water from the aquifer each year. As many as 6,000 wells have been sunk into the aquifer, many of them without authorization. Many are drilling the illegal wells in the middle of the night to avoid prosecution. Water officials fear the aquifer could become unusable within just a few more years, leaving Gaza's population of 1.6 million without water, and with the potential to greatly increase political tensions with neighboring countries.

Effective governmental or communal control on water use is absolutely essential in averting tragedy in the water commons. An effective water allocation process can be guided by governmental regulations or policies or by cultural norms and community forums. Unfortunately, very few states, provinces, countries, or local communities have put adequate controls in place, even when it is obvious that a water source is being exploited in an unsustainable manner. For instance, very few governments are regulating groundwater use effectively, and as a result, many countries are extracting groundwater faster than it can be replenished: Mexico's consumption of groundwater exceeds natural recharge by 20 percent, China's by 25 percent, and India's by 56 percent.[4] In some parts of China, the groundwater level has dropped 50 meters (165 feet) in the past 50 years, and continues to drop by 5 meters every year. Since 1900, the United States has drained its aquifers by a volume equal to two Lake Eries.[5]

The tragedies of the water commons unfolding around the planet today have seldom resulted from a lack of rules for allocating water, however. As discussed below, rules or customs for allocating or sharing water exist in most parts of the world. Water shortages are instead caused by ineffective implementation or inadequate societal adherence to existing rules.

Different Approaches for Allocating Water

Governments use a variety of procedures, rules, or customs for issuing entitlements to use water, and it is not uncommon for more than one water allocation process to exist within a political jurisdiction, particularly in the case of different rules for surface water and groundwater. Some types of water use are assumed to be intrinsic or granted automatically, such as when a government asserts that each of its citizens has a right to sufficient water to meet the basic needs for drinking, cooking, or washing. Of important note here is the International Covenant on Economic, Social and Cultural Rights, a multilateral treaty adopted by the United Nations General Assembly in 1966 pledging commitment to basic rights for citizens of signatory countries. In its interpretation of this covenant, the United Nations has clarified that "the human right to water entitles everyone to sufficient, safe, acceptable, physically accessible and affordable water for personal and domestic uses." As of 2013, the covenant had 160 signatory nations, giving notable weight to the basic human right to water. The intrinsic right to use water does not mean that everyone has been provided adequate access to it, however, as evidenced by the fact that nearly 800 million people—more than a tenth of all people on the planet—still lack access to clean drinking water.[6] Even when governments are morally committed to providing such access, they often have not been able to adequately deliver water to all who need it, for reasons detailed later.

Many governments continue to recognize communal or traditional processes for allocating water, by adopting tribal or customary laws or rights that predate the formation of the existing government or legal systems, for example. Much of this community-based law is informal and unwritten, but it guides the water-use behaviors of hundreds of millions of rural water users in developing regions. Anchored in the wisdom of time and place, these community-based laws have proven to be quite robust and durable, and most importantly, they reflect the local cultures and belief systems of community members.[7]

Many rural cultures, for example, believe in the existence of deities living within freshwater sources. Such beliefs have been very important in establishing cultural norms for the use of the water. The people of Besao in the northern Philippines, for example, believe that the *nakinbaey* is a supernatural being inhabiting water sources and other sacred sites in their region.[8] Water is believed to be produced by the nakinbaey. To ensure adequate water supply, therefore, people should do nothing that would upset the nakinbaey and cause it to leave the source. This requires people to consciously respect the water source by observing behavioral norms that strongly encourage water sharing within the community and help to protect the quality of the water source, such as prohibiting grazing animals in the vicinity of water sources, as the breath and waste of cattle are believed to be repulsive to the nakinbaey.

In many countries or local jurisdictions, the right to use water is tied to the ownership or use of land. One common example of this approach is a "riparian right," which is an entitlement given to landowners located along the bank of a river or lake. Similarly, landowners may be entitled to use the groundwater beneath their land, or the rainwater falling on their land, oftentimes without restriction.

In drier regions, including many places now experiencing water shortages, the most widespread process for allocating water is an administrative process in which some governmental entity, such as a water resource agency or irrigation board, is given the authority to issue water entitlements, which grant the right to use water. These entitlements to use water will usually come with certain attributes or rules that specify the terms or limits of the water entitlement. Some examples of these attributes are presented in table 4-1.

What Is Going Wrong?

The existence of rules for allocating or sharing water, whether formally institutionalized in national legal systems or based on unwritten community law, has not stemmed the depletion of the planet's water sources, unfortunately. These rules have been ineffective for a number of reasons.

Table 4-1 Examples of Typical Attributes of a Water Entitlement

Quantity:	The amount of water the holder of the entitlement may withdraw or consumptively use, or the area of land and crops that can be irrigated
Quality:	The quality of the water to be withdrawn or disposed of
Source:	The specific source and location from which the entitlement is awarded
Timing:	Restrictions on the time when the entitlement applies, that is, times when the volume may be withdrawn or consumptively used
Assurance:	Absolute entitlement, meaning the volume is always fulfilled, or variable assurance of supply depending on how much water is available each year
Type of Use:	The specific use for which the water is to be withdrawn or consumptively used (e.g., irrigation, mining)
Duration:	Permanent entitlement or entitlement authorized for only a specified period of time
Transfer:	Whether the entitlement can be sold, transferred to another person or location, or inherited

Adapted from Tom Le Quesne, Guy Pegram, and Constantin Von Der Heyden, *Allocating Scarce Water: A WWF Primer on Water Allocation, Water Rights and Water Markets* (London: WWF, 2007).

Insufficient Financial Capacity

The most pervasive problem in governments is a shortage of funds. Money is needed to pay the water managers that work for the agencies that issue and administer water entitlements. Money is needed to build and maintain the infrastructure to store and distribute water to users. Money is needed to pay for the computers that are used to create models of watersheds and aquifers and to store the data that help water managers keep track of water entitlements and violations of those rights. Money is needed to operate scientific instruments and gauges that keep track of water availability and use. If the government is not generating enough money from taxes or other means to meet these needs, or if the government simply is not allocating enough of its resources to the governing of water, water allocation and management systems are almost certain to fail.

The United States builds, operates, and maintains some of the most sophisticated water management systems in the world. But two disturbing trends illustrate what can happen when insufficient funds are dedicated to managing water. One trend relates to the fact that urban water supply systems in the United States have not been properly maintained. Many of the reservoirs, pipes, and water treatment facilities found in U.S. cities were built more than 100 years ago. That infrastructure is now desperately in need of repair or replacement. Due to decades of neglect, upgrading the nation's water and wastewater systems is now expected to cost nearly $384 billion over the next 20 years.[9]

Another distressing trend in the United States is the loss of river monitoring stations. Since 1980, more than 2,300 river gauges—representing more than one-fourth of the monitoring network maintained by the U.S. Geological Survey—have been shut down because of inadequate funding.[10]

In India, in an effort to give farmers greater responsibility for infrastructure management and its cost, the national government is moving toward "participatory irrigation management," including handing over responsibility for management of irrigation systems to local water user associations. The government's water resource department is supposed to do a one-time rehabilitation of the infrastructure before transferring ownership to the local associations, but this often does not happen, as the government insists there is no money to do so. As a consequence, the farmers are often handed a system in terribly degraded condition.

The deteriorating state of water infrastructure in many countries is not entirely the fault of the federal or state governments or their failure to allocate sufficient revenues to these purposes. Public water supply systems have been deteriorating because the local water agencies that own and maintain these systems either have not properly planned and budgeted for necessary maintenance of their infrastructure, or have been unable to charge enough for their water deliveries to cover these costs. Many people believe that in keeping with the philosophy that access to water is a basic human right, water should be delivered free of charge. That sentiment makes it politically difficult for water managers to generate sufficient revenues from water sales to keep water systems running properly. While water costs should be subsidized or kept low for poor sectors of our society to the extent possible, we should each understand that we need to pay for the services required to manage our water supplies.

Lack of Expertise

Another problem related to money is a lack of expertise to manage water well. Engineers and hydrologists are needed to design and maintain reservoirs; water distribution systems; facilities to clean, recycle, or desalt water; and other types of water infrastructure. Technical experts also design and operate monitoring systems to keep track of water availability and use. Financial experts and accountants are needed to set water prices, finance water projects, send out water bills, pay contractors, and manage budgets. Environmental scientists are needed to design strategies for managing the health of watersheds, or the proper ecological functioning of freshwater ecosystems. Social scientists or negotiators are needed to help facilitate discussions and negotiations among water users, politicians, and various government agencies. Lawyers and policy makers are needed to set rules for water governance, prosecute violators, and mediate conflicts. The best-run water agencies in the world possess most or all of this expertise, employ hundreds of staff members, and operate with multimillion-dollar budgets. But possessing such capacity is the exception and not the rule.

Lack of Will to Enforce Rules

Enforcement of rules is almost always the weakest link in any system for managing water. The day-to-day management of water, particularly in rural areas, takes place in absence of any monitoring systems and without oversight of government officials. That means that friends, peers, and neighbors play a default role of discouraging bad behavior or breaking of rules. It can be extremely difficult for neighbors or family members to hold the line on water abuse in local communities, particularly when the enforcement of rules has implications for livelihoods, incomes, or family relationships. On the other hand, peer pressure and social norms or morals can be much more powerful and effective than any formal law in controlling against cheating or abuse, particularly when no government official is watching.

Lack of Coordination among Authorities

There are also serious challenges in coordinating the diverse areas of expertise and governmental units responsible for different aspects of water management. As governments grow, shrink, and evolve in response to funding availability and changing political priorities, it is very easy for different governmental departments to become isolated into "silos" that do not communicate or coordinate well with each other. In most countries, for example, different units of government have been set up to manage agriculture, mining, forestry, energy, environmental quality, water, and other resources.

When those ministries, agencies, or departments do not talk and plan well together, bad water policy commonly results. In India, state governments heavily subsidize the use of electricity as a social service to help poor farmers, but these subsidies have wreaked havoc on the financial viability of electricity utilities in the country and have made it extremely difficult for local water managers to control the overpumping of groundwater because the electricity costs very little or nothing. Similarly, government subsidies on diesel oil have enabled farmers outside the capital city of Sana'a in Yemen to greatly expand their use of groundwater because they can run their well pumps cheaply, particularly for growing khat, a narcotic that requires extremely water-intensive cultivation. As a result, the Yemenis are rapidly depleting both the regional aquifer, which sustains the farms and the capital city, and the oil reserves that provide three-quarters of the government's revenue. In the United States, the Department of Energy helped create a federal mandate for increasing biofuel production in the interest of reducing carbon emissions that drive climate change. Similarly, more than sixty other countries have now created mandates or subsidies to stimulate biofuel production.[11] Those subsidies, however, have substantially shifted overall agricultural production toward biofuel crops such as corn for ethanol, causing a spike in consumptive use of water due to greatly increased irrigation needs for biofuels, and generating widespread shortages of corn and other foods for human consumption.[12] In the interest of bolstering its national gross domestic product with massive infrastructure construction projects, the State Council in China has repeatedly funded some of the world's largest dams and water diversion projects, even when local water agencies have identified more cost-effective measures with lesser environmental and social impacts. Given the difficulties of designing and implementing good water policy within any single country, imagine the challenge of creating a sensible water policy or treaty among multiple countries!

Even within ministries or agencies directly responsible for water, there may be multiple subunits that can become quite insular and disconnected. For example, one agency might be responsible for surface water while another agency is responsible for groundwater. One agency might be responsible for water quantity and issuing entitlements for water use, but another is responsible for water quality and punishing polluters. One department might be responsible for building, operating, and maintaining water infrastructure, while another is responsible for installing and maintaining the monitoring systems that tell us how well that infrastructure is working. Not surprisingly, virtually every government struggles with conflicting authorities, duplicative and redundant responsibilities, and management inefficiencies.

Too Much Corruption

The saddest failures in governing water are caused by corruption. Transparency International, an international organization that monitors corruption, has repeatedly found the water development industry to be one of the most corrupt economic sectors. One of the more widely publicized examples involved Masupha Sole, former chief executive officer of the Lesotho Highlands Development Authority in Africa. Sole was found guilty in 2002 of accepting more than $6 million in bribes from engineering and construction companies. Until he was sentenced to 18 years in prison, Sole was managing an $8 billion water project involving the construction of five dams and multiple large canals for water storage and transfer, hydroelectric power generation, and rural development in southern Africa. The court case marked the culmination of an investigation that unveiled a complex and intricate web of corruption in the water industry, implicating twelve major multinational firms.

The impact of corruption is well explained in Transparency International's *Global Corruption Report 2008*, by Kristen Lewis and Roberto Benton. They cautioned, "The lure of milking big ticket projects for private gain may keep officials from exploring a wider range of alternatives, such as water conservation. In particular, corrupt decision-makers may favor projects where corruption payments are concentrated, and can be easily appropriated by them or their chosen cronies, over smaller projects, which disperse corruption payments more widely."

One of the biggest criticisms directed at the Lesotho Highlands Water Project by its detractors was their claim that water conservation could have met all water needs for 12 to 15 years, thereby postponing if not wholly obviating the need to increase water supplies with a new water diversion. Instead, the project diverts 40 percent of the average flow of the Senqu River, moving it through hundreds of kilometers of pipes and canals to the Guateng Province in South Africa.

Water corruption takes many forms, most of which are far more subtle than outright bribery. Dan McCool, writing in his book *River Republic*,[13] explained the role of political manipulation in U.S. water development projects in the twentieth century: "It did not take long for both the Corps [of Engineers] and the Congress to realize that some form of Corps project, paid for by the taxpayers of America, could generate a lot of votes and contributions for a legislator's next campaign. Water projects would help a lot of legislators get elected—again and again and again. Projects became a kind of political currency, to be traded in the halls of Congress for favors and votes. That, in a nutshell, is why we have so many dams, levees, channels, and waterways. The projects were sometimes in the national interest, occasionally in accord with sound economic principles, but rarely built in an environmentally sound manner, and sometimes a gross waste of money."

In India, this is called "vote bank politics." As K. J. Joy, a water researcher and activist in the country puts it, "Water is a vote catcher. All political parties and politicians promise water and free or cheap electricity to pump water as part of their populist promises to get votes. This has played havoc in the water sector, leading to overpumping of groundwater and failed attempts to extend surface water projects beyond their capacities."

What Can We Do to Improve Water Governance?

The legacy of Thomas Jefferson—the third American president and principal author of the nation's *Declaration of Independence* from British rule—is abundantly evident where I live, in Charlottesville, Virginia. Jefferson's beautiful Monticello home sits upon a hilltop that overlooks the city and the University of Virginia, which Jefferson founded and designed, and where I teach a water course. Jefferson was an eloquent spokesman for the rights and roles of everyday citizens, and he believed fervently that citizens play an absolutely essential role in shaping their governments. He was also very cognizant that citizens must be well informed if they are to effectively control their destinies. Jefferson put it: "An informed citizenry is the only true repository of the public will."

It is my sincere hope that this book can help those interested in or concerned about water to become more informed about options available to them, and stimulate much greater citizen involvement in water decision making. It will require more than an informed citizenry, however, to enable broader engagement of water users and interested citizens in water allocation, planning, and management. Governments must be willing to invite and embrace such engagement in decisions that have long resided in their exclusive domain. Fortunately, a strong wind of change is blowing through many of the halls of governments around the world, creating new opportunities for individuals, communities, and businesses to help resolve our water crisis.

Water Is Everyone's Responsibility

The concept of a water governance system, if implemented well, offers considerable hope and opportunity for citizens and private businesses to become more directly involved in their water management. Rather than viewing governments as the sole arbiters of water, a water governance approach can greatly expand the roles and responsibilities for allocating and regulating water, and balancing water budgets. It is very important to understand, however, that governance structures and functions will likely need to be custom tailored to local regions and even to individual water sources. And it is also important to understand that it will rarely be advantageous or desirable to supplant the primary roles of government, such as water allocation and regulation. Instead, new participants in water governance should look for ways to strengthen essential governmental functions.

The pursuit of effective water governance is in essence a search for the right balance in authorities, leadership, and resources among governmental entities, the private sector, and civil society. The private sector includes private businesses and corporations of all sorts, many of which necessarily use a lot of water, such as in energy generation, mining, manufacturing, and agriculture. Civil society refers to individuals and organizations concerned with water, ranging from local watershed activists to subsistence farmers, and also including nongovernmental organizations that represent environmental or social interests.

Implementing good water governance will require inviting and balancing the input and influence of public and private interests. This does not mean that each group should be deeply involved in all instances. In India and a number of other countries, for example, there is great concern about the influence of private companies in water affairs. The "right"

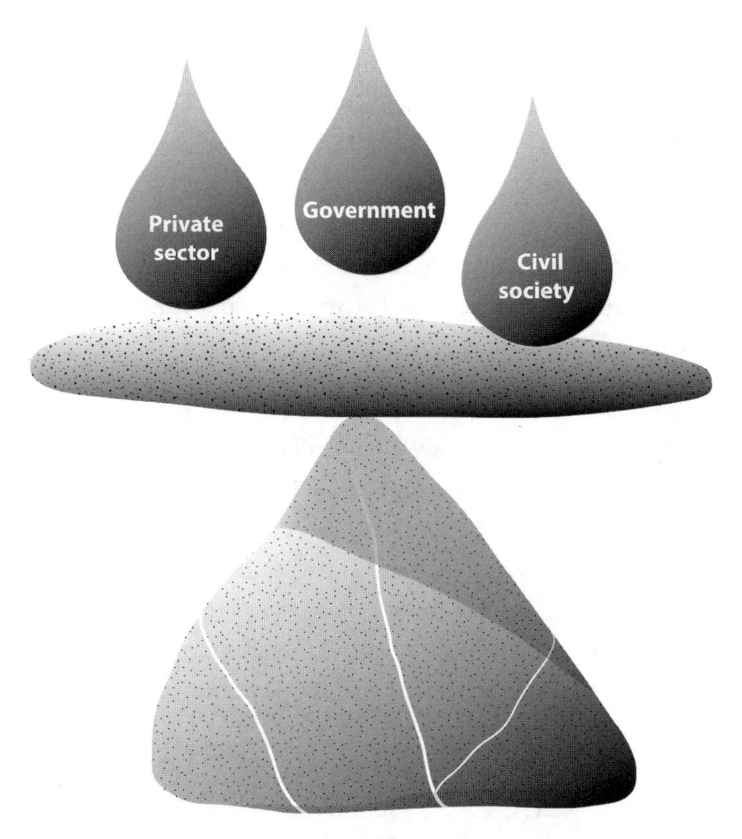

Figure 4-1. *Water governance can be envisioned as a balancing act involving government, the private sector, and civil society. Getting governance right is like trying to balance a river stone, with each governance sector possessing different weight in decision making, threatening to imbalance the river stone. When things are not going well, it can be helpful to ask whether each sector's influence is weighted properly.*

balance, then, is the balance that is most appropriate for the culture, laws, and economies of each place.

The process of water governance is a lot like trying to balance one river stone atop another, as in figure 4-1. Finding the right balance point will need to be explored, adjusted, and set for each stone. I have found this stone-balancing metaphor to be quite useful in my own work. When I begin to become involved in the water issues of a new place, I will often start by assessing the balance, or imbalance, among the participants. Is the government's capacity strong enough, or is it sufficiently receptive to external engagement? Are private companies exerting too much influence on decision making or causing problems through their use of water, or can their resources be used to beneficial effect? Do concerned citizens and nongovernmental organizations have sufficient access to voice their opinions effectively, and are they sufficiently educated on water issues to make well-informed contributions?

As citizens and private interests become more engaged in water decision making and governance, there will be many diverse functions and roles to be examined, adjusted, and strengthened. Only the most common needs are addressed here.

Ensure Adequate and Stable Funding

First, it is very important that we financially support our government's efforts to manage our water resources well. As discussed earlier in this chapter, there are many expenses associated with proper management and governance of the water that we all use, and we all need to help pay for these essential services. There are many ways to generate public revenues for this purpose, but emphasis should be given to funding strategies that can best endure downturns in the general economy, and can weather changes in political administrations.

The state of Florida in the United States offers one cautionary example. The state's five water management districts have long been heralded as being among the best examples of well-run government agencies. With relatively stable financial support that comes from a tax on real estate property, the Florida water districts have become widely recognized leaders in watershed protection, environmental restoration, water conservation, climate and water monitoring, and preservation of the ecological health of the state's water sources, even while enabling strong economic growth. However, when a new state governor was elected in 2011, he mandated deep cuts in the water district budgets in an effort to reduce the tax burden on the state's residents during a difficult economic downturn, resulting in much-reduced capacity for water data collection, research, and other efforts. While no public agency budget is immune to budgetary cutbacks, and in many cases such budget adjustments may be supported by the general populace, it is very important to heed Jefferson's urgings to continually educate and inform our fellow citizens so that they understand what they will lose when budgetary reductions are executed. As Jefferson suggested, that education will enable them to participate most productively in governance.

Plan for the Long Term

Another very important aspect of good water governance is long-range planning. Many water agencies or governance groups have adopted a 50-year planning horizon for their water management, which provides important opportunities to anticipate future water needs and potential changes in water availability due to climate projections, and to thoughtfully plan and budget for activities and water infrastructure to address the water needs of coming decades. A long planning horizon can help to smooth out the costs of needed infrastructure or other large expenditures, as opposed to having to impose abrupt increases in water prices to pay for such investments. While it makes great sense to plan far ahead, it is also important to revisit those long-range plans regularly so that changes in projections can be integrated into the plans. For example, many water planning agencies or groups adjust their 50-year water plans on regular intervals, such as every 5 years.

Find Ways to Better Coordinate

There is no fail-proof recipe for ensuring that the individuals and departments in a large and complicated government system will communicate and coordinate their plans adequately. There is no one perfect organizational structure of ministries, agencies, or departments that will ensure that all of the important contributors will always get together to talk about the right topics at just the right times, or align their actions effectively. However, case studies from around the world suggest that a watershed or aquifer can be an extremely helpful hub for organizing water dialogues, helping to coordinate and focus discussions and action on water issues pertinent to a specific water source. Numerous examples exist of well-functioning watershed councils, aquifer management districts, and river basin organizations whose focus is a single water source. These organizations can be set up by political authorities, or they might instead be formed by concerned local citizens. Regardless of origin, the important feature of these water-based cooperatives is that they are organized around the issues pertinent to a particular water source.

A primary challenge is deciding what scale will be most effective in coordinating activities and decisions. Imagine the challenge of planning for the water future of a huge watershed like the Colorado or the Brahmaputra! As a general rule, the dialogue circle should be kept as small as possible and as big as necessary to address the problems at hand. For example, a local community may find it desirable to appoint a group of community members to lead discussions about a small watershed of a few thousand hectares that provides their water supply. At the other end of the spectrum, concerns over pollution in the Black Sea of southeastern Europe brought seventeen countries together to develop the international agreements necessary to address their problems. Sometimes it will be helpful to develop a nested system of planning groups, with some groups focused on issues within subwatersheds and another group responsible for overall coordination and planning at the whole-watershed scale.

The choice of scale for resolving water issues should always be made based on the questions and problems being addressed. Some problems are best resolved by neighbors over a cup of tea, but other problems can only be deliberated by high-level representatives of major economic sectors, interest groups, or governments. The common threads running through all successful and durable water negotiations are respect, trust, and a feeling of ownership of the process. Sharing or governing water will work best when we understand what each of us wants and needs. And that can only happen when people talk to each other, and learn from each other.

Learn from Others and Emulate the Practices That Work Well

There are literally thousands of local water sources on our planet. There is a great deal to be learned from both the mistakes and the successes in managing those watersheds and aquifers to date. In the next chapter, I will offer seven sustainability principles that have led to success in many different places. You might find it useful to use those principles as a scorecard of sorts, to evaluate how well your local water source is being governed. That assessment can also help to identify weaknesses in the current water management system that can become the foci of community-based, collective action to improve water governance. In chapter 6, I will offer some examples of water user groups that are working toward improving the governance of their water sources.

CHAPTER FIVE

Seven Principles for Sustainability

"People used to think that it was a curse from God. Some people thought that they were bewitched. But later we came to realize that it was those dams. The dams are swallowing a lot of water. Now water cannot come here."

Chief Omar Abdalla Hama was describing the terrible hardship his Ozi Village tribe has been through since five large dams were built far upstream on the Tana River in eastern Kenya. During my visit with Nature Conservancy colleagues in 2011, Hama pleaded with us to help save his people from starving. As he walked around the *shamas*, or farm fields, within his village, Hama pointed at the withering rice plants that no longer received the freshwater overflow from river floods that they needed to grow. The dams had shut down the natural flooding process. Hama told stories of many community members who had fled the village to seek food in other places.

The dams that Hama was complaining about were built in the late 1970s and early 1980s to capture the rainy season floods on the Tana River and put the water to urban uses. The water and the electricity generated by the dams are of critical importance to the capital city of Nairobi and other, smaller towns. But now the river can no longer adequately support the hundreds of thousands of people living along the river, whose livelihoods and survival were intimately tied to the river's natural flow.

In their free-flowing form, large rivers like the Tana are among the most productive, life-supporting ecosystems on the planet. These natural supermarkets continue to feed hundreds of millions of poor people each and every day. River and floodplain fisheries are a critical source of food and income for at least a billion people living in developing regions. For example, the Mekong River's fish are the primary source of protein for 60 million people.

Things did not have to turn out badly for Hama and his Ozi Village. If the dams had been designed and operated differently, they could have continued to release a yearly flood or two sufficient to sustain the fisheries and floodplain crops along the lower Tana River and its delta, even while providing electricity and water for cities. But the hundreds of thousands of river-dependent people living along the river were not consulted about the dam and its potential impacts, and their needs were not considered when the dams were built.

Water resources cannot be managed in a fair, equitable, and sustainable manner without providing all interested and affected parties with an opportunity to express their values and needs. Building a shared vision for the use of a water source is one of seven "sustainability principles" that I will offer in this chapter (table 5-1). Applying these seven principles will not guarantee success in water management, but evidence from thousands of watersheds and aquifers around the world suggests that when one or more of these principles are neglected, the potential for social conflict can rise substantially. There is also abundant evidence that each of these principles can be of great benefit when applied well.

A Framework for Managing Water

In chapter 3, I described six water tools that can be used to rebalance a water account when it is being overtaxed. But no tool can be used effectively without a plan for what you are going to build with it. None of the water tools discussed previously will prove helpful over the long term absent a water governance framework that guides the use of those tools. In this chapter I offer the beginnings of a framework for building a sustainable water program, in the form of seven guiding principles that should be helpful to communities and governments wanting to move along the pathway toward water sustainability. There are many approaches to governing water that have been successful even when omitting some of the components described here, but each of these principles can help ensure success.

Table 5-1 Seven Principles for Sustainable Water Management

Principle #1:	Build a shared vision for your community's water future.
Principle #2:	Set limits on total consumptive use of water.
Principle #3:	Allocate a specific volume to each user, then monitor and enforce.
Principle #4:	Invest in water conservation to its maximum potential.
Principle #5:	Enable trading of water entitlements.
Principle #6:	If too much water is being consumptively used, subsidize reductions in consumption.
Principle #7:	Learn from mistakes or better ideas, and adjust as you go.

Principle #1: Build a Shared Vision for Your Community's Water Future

In my work with community groups, I always start by asking them what they think good water management would look like. Invariably, many of the respondents will say that they want their water sources to be managed "sustainably." Yet most of us struggle to define sustainability, or translate it into an action plan. Peter Gleick, a widely respected water policy analyst, has offered a definition of sustainable water use that offers a great starting point: "the use of water that supports the ability of human society to endure and flourish into the indefinite future without undermining the integrity of the hydrological cycle or the ecological systems that depend on it."[1]

While most communities and governments would likely embrace this aspiration, very few are earning high marks when measured against it. I attribute much of this failure—as so well illustrated by Syria's plight and the Tana River story above—to the fact that governments do not ask water users what they need or want, or facilitate a dialogue that could lead to a shared vision within the community, a problem addressed in the last chapter. When a shared water source is used too heavily or water users are not getting what they want or need, conflict is certain to arise.

Developing a shared vision for water is no small challenge, however. There are many competing demands being placed on the planet's water resources these days. All members of a community want a sufficient supply of clean water to meet their basic needs for drinking, cooking, and washing. Some individuals and their families may want enough water to grow their own food, perhaps growing enough to take a surplus to market, or they may want to leave enough water in their local river to support productive fisheries. Some may want water to grow flowers or a lawn to beautify their property. Some may want enough water to be present in a river or lake to enable their recreation, such as boating, or to transport their goods to markets. Some may strongly value the presence of flowing water in a river for aesthetic reasons, or because their religion or spirituality is tied to flowing water, or because they want natural ecosystems and biodiversity to be sustained.

Public or private companies may want to use water to provide services or goods, such as by supplying drinking water or electricity to your community. Other companies want to use water to produce consumer goods to sell for profit—farmers produce and sell crops and meats, and manufacturers and industrialists make the products used to build our homes and businesses or fill them with things that support our lifestyles or livelihoods.

All of these pursuits require the withdrawal and consumptive use of water. Is it any wonder that when our water sources are strained, people will hold different opinions and perspectives about how water should be used and managed?

Effective and equitable facilitation of local dialogues about water sharing is likely to become the greatest social challenge of the twenty-first century in water-short regions of the globe. In the next chapter I will discuss some social processes that appear to be succeeding in bringing community members and water users together to discuss their needs and values. The ability to convene, facilitate, and sustain an open dialogue about values and needs is essential to sound water governance. Water user communities will need to openly and explicitly address difficult questions: Is maximizing economic productivity and jobs our highest priority? How much do we value social equity, that is, managing water in ways that allow all members of our community to meet their basic water needs, even perhaps have access to enough water to generate income, or simply to be able to afford the water utility bill each month? How much water do we want to leave in our local rivers to support fisheries, provide recreational opportunities, sustain the beneficial services that ecosystems provide, or support aesthetic or spiritual values?

Using and managing water in a way that an entire community of water users might feel is optimal or sustainable is extremely difficult. But we have to try our best to find ways of sharing water that are fair, equitable, and mutually acceptable. There is only so much of it to go around, only so much that any community can reasonably access. As Kevin Rogers of South Africa once said to me about the challenges his country faced as it emerged from apartheid in the mid-1990s, "In our allocation of water we will see what democracy will mean for our country." All water-scarce regions of the world face this same challenge of water governance.

How Much Water Use Is Too Much?

From our evaluations of water scarcity and its impacts on local communities, economies, and ecosystems, my colleagues and I have identified a couple of thresholds in water use that might be helpful to communities in their deliberations over how to use their available water. These thresholds are not universal, nor are they fixed and inflexible, but some general rules of thumb can be quite instructive.

First, mounting evidence from around the globe suggests that when the day-to-day water flow in a river is depleted by more than about 20 percent, it becomes increasingly likely that the ecological health of a river—the river ecosystem—will suffer.[2] What that means is that populations of aquatic species such as fish may begin to decline, some sensitive species may disappear altogether, or a river's ability to provide important ecological functions such as flushing away waste material or other pollutants will be increasingly diminished. Of considerable concern is the fact that more than half of the world's rivers are being depleted by more than 20 percent during some portion of the year, which helps explain why freshwater animals such as fish, turtles, and frogs are the most imperiled groups of species on our planet.[3]

It appears that ecological thresholds for aquifers and lakes may be even more sensitive than for rivers. These water sources present a rather ironic quandary: they may hold tremendous volumes of water in storage, but even small drops in their water levels can cause ecological damage. For example, in lake ecosystems many aquatic plants and animals and many important ecological processes are strongly dependent upon the presence of wetlands and other shallow habitats formed along the lake's shallow fringe. A drop in the lake level of as little as a meter, or even less, can dry out and disrupt those habitats and processes, with great consequence for the productivity of the overall lake ecosystem. Similarly, many shallow aquifers continually drain into rivers and springs, providing reliable and critically important (and usually cooler) water flows during dry seasons and droughts. When aquifer levels are lowered because of excessive pumping, the flow of water from the aquifer into rivers and streams can disappear.

While these general rules of thumb can provide useful indications of the level of water depletion that may cause ecological harm, water planning efforts should include, whenever possible, further investigation into the ecological sensitivities of the particular freshwater ecosystems that could be affected by water use. A scientifically based "environmental flow assessment" can help determine the volume and timing of water flows needed to sustain ecological health and other social values. A variety of approaches can be employed for this purpose, at varying expense and time commitment. By investing in an environmental flow assessment, water planners and communities of water users will gain a better understanding of the species or ecosystem functions that might be affected at different levels of water depletion.

Similarly, there are no absolute or universal thresholds of water depletion at which economic productivity will be jeopardized. But as discussed in chapter 1, it appears that when more than half of the available, renewable water supply is being depleted on an ongoing basis, the community of water users will likely face a serious risk of running out of water during periods with less-than-average rainfall or snow. When water shortages hit, they can have severe economic impacts.

Along the gradient running from 0 to 100 percent depletion of a water source, there exists a long continuum of options for the management of a water account. There is no golden rule setting out an optimal level of water use or depletion—that decision must be undertaken by the community sharing a given water source, ideally through an effective water governance approach, as discussed in chapter 4. The community must balance ecological protection and basic human needs for water with economic productivity. These decisions should be perpetually revisited, at regular intervals or during times of crisis, as the needs and values of water users and other citizens change over time.

Build a Shared Vision as Early as Possible

While there are many compelling reasons for developing a shared vision for managing local water sources before those supplies are stressed or a crisis develops, it seems that one of humanity's greatest failings is our inability to use the information available to us to plan and act with foresight. Far more commonly, we proceed down a path without a plan, until we encounter trouble. Only then do we take corrective action. As any individual or government that has experienced bankruptcy knows, it is much easier to pull back on spending gradually and incrementally, rather than implement severe and painful budget cutbacks during a fiscal crisis.

A telling example is the Tablas de Daimiel National Park in Spain, which includes more than 2,000 hectares (5,000 acres) of wetlands along the Guadiana River. The shallow groundwater aquifer underlying the park is recharged by both rainfall and seepage from the Guadiana River. However, the flow of the Guadiana has been severely depleted in recent decades by agricultural water consumption, resulting in much-reduced recharge of the groundwater underlying the park. Additionally, farmers have dug many groundwater wells around the park's edges. Consequently, the groundwater level has dropped by more than 20 meters in the park, desiccating the wetlands.

When temperatures soared during the summer of 2009, the dried organic matter in the wetlands spontaneously combusted and caught fire. Photos of this raging wildfire showed up in newspapers and on TV channels across western Europe. Media reporters pointed to the catastrophe as an example of "what not to do" in water management. Tourism in the park came to a halt, affecting the economy of local communities.

Concerned over ecological impacts—the wetlands in the park have been designated as being of international significance—the European Union gave the Spanish government 10 weeks to explain how they would fix the problem. Unable to immediately reduce water extractions from the river and aquifer, government officials decided to pipe water 150 kilometers (100 miles) from the Tangus River to temporarily refill the park—at considerable expense. The Spanish government and local communities now realize that they will need to aggressively reduce consumptive use of water from the river and aquifer if they want to continue to enjoy the tourist revenue and other benefits from the park.

These painful economic, social, and ecological disruptions could have been avoided by developing a water budget for the river and aquifer, facilitating a community dialogue about water use, and putting into place adequate water rules commensurate with the community's vision.

Principle #2: Set Limits on Total Consumptive Use of Water

Creating a shared vision for the management of a water source—as discussed under principle #1—is a highly desirable precursor for establishing practical rules to help realize that vision. The single most important rule that a community of water users can adopt will be the setting of a limit or "cap" on total consumptive use of water. Without such limits, it is very likely that depletion of a water source will eventually progress to the point where highly undesirable impacts begin to appear.

Establishing a cap on the total volume of water that can be consumed from a particular water source does not necessarily mean that new uses of water must be precluded once water use reaches the cap level. If existing users of water become more efficient in their use, which means using less water to serve the same purpose, the saved water can be made available for new uses. For instance, more residents can be added to a growing city without causing an increase in the total volume of water use if everyone in the city is able to reduce the volume of water used each day. Similarly, more crops can be grown without an increase in total consumption if all farmers are able to irrigate more efficiently. Alternatively, additional or new water supply can be provided by using one or more of the water supply tools described in chapter 3, such as by importing water from another water source or by desalting ocean water, thereby avoiding the need to increase use of local freshwater sources.

There are many different ways to limit or control water use through regulations or incentives. Three widely used examples, each based on issuance of water entitlements as discussed in chapter 4 (e.g., see table 4-1), are described here.

Option 1: Create a Virtual Lineup of Water Use Entitlements, and Supply Water to Each Sequential User in the Line until All of the Available Water Has Been Consumed

For example, in the western United States, many state governments have adopted an approach for issuing water entitlements that is based on the concept of "first in time, first in right," also known as the "prior appropriation doctrine." Using this approach, water users are (virtually, not literally!) lined up according to the date at which each first started using water; this timing of first use sets the "priority date" that is assigned to each water entitlement. The volume of water use authorized by each individual entitlement is predicated on the amount of water used originally. At any given time, each water user in the priority line is sequentially granted the full volume of their water entitlement until all of the available water is used up. This does not necessarily mean that each sequential water user is served until the water source is dried up completely. For instance, if a decision has been made to always leave some volume of water in a river or lake to protect its ecological health, the priority line will be shut off when only that environmental water remains in the source.

This prior-appropriation system has certainly provided a useful means of clarifying rights to use water, but time has revealed some shortcomings in this approach. It can be very difficult to know how many of the water users in the line can be served at any time, because it can be very hard to keep track of exactly how much water is actually available from day to day or month to month. It is also very tough to guarantee that shutting off a "junior" water entitlement holder in one part of a watershed will ensure that enough water remains to fulfill the entitlements of more "senior" water users in other parts of the watershed, given the huge challenge of knowing when and where every water withdrawal, return flow, and consumptive loss is taking place at any given time. Perhaps the greatest shortcoming of prior-appropriation systems is the fact that—as so well illustrated by the Colorado River story in chapter 1—freshwater ecosystems and species were not given a place in the lineup of water entitlements until very recently, meaning that in times of shortage, nature is often the first to lose.

Option 2: Create a Reserve of Water to Ensure That Basic Human Needs and Ecosystems Are Protected, Then Allocate the Remainder

For example, South Africa is presently establishing a water reserve in each of the country's watersheds. These watershed-specific reserves have two parts: a basic human needs reserve, which helps to ensure that everyone living in the watershed is guaranteed sufficient water to meet basic needs such as drinking, cooking, and washing; and an ecological reserve, which is intended to leave enough water in rivers or lakes to support their ecological health. Once this two-part reserve is quantified for each freshwater source, the remaining water supplies can be allocated in the form of entitlements for other uses. The national water act that institutionalized this concept of water reserves in South Africa is widely recognized as a major breakthrough in water policy because it protects both basic human needs and the ecosystems that many South Africans depend upon for their livelihoods and food security.

Implementation of this reserve approach has suffered from many of the same challenges mentioned above for the prior-appropriation approach. In particular, it has been very difficult to know how much water can be allocated to entitlements, because the ecological needs do not remain constant; in other words, the ecological reserve is a highly variable reserve. Many of the water prescriptions prepared by the country's river scientists in an effort to quantify the ecological reserves contain very detailed specifications for maintaining different river levels at different times of the year, or during droughts or high-water periods, to provide desired habitat conditions or ecological functions. These fluctuating ecological requirements have frustrated the country's water managers, who must try to continuously meet those ecological targets while at the same time fulfilling the water entitlements of other users.

Option 3: Set a Cap on the Total Volume of Water That Can Be Consumptively Used during the Driest Years, but Then Allow Additional Allocations during Wetter Years

The Australians have implemented a variation of this approach in the Murray-Darling watershed by creating two types of water entitlement: a high-security entitlement that comes with a specified volume of water that is intended to be fully served in virtually all years; and a low-security entitlement, the volume of which can be adjusted each year according to available water supplies. The total volume of water that is allocated to the sum of these entitlements is limited in a way that ensures that enough water will be available in the river system to protect ecological health.

In deciding how much water to allocate to both high- and low-security entitlements, the Australians—like the South Africans—rely quite heavily on scientific analysis of ecosystem needs in determining how much water must be left in their rivers to keep them healthy. However, there is a subtle but very important difference in the way that water is provided for ecological purposes in Australia. Rather than focusing on maintaining

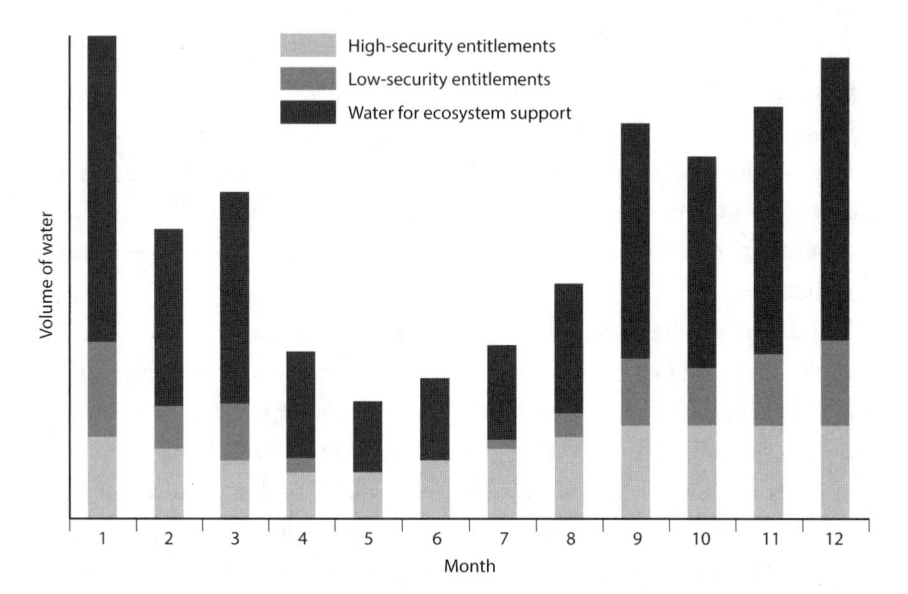

Figure 5-1. *This chart illustrates the concept of cap-and-flex for one hypothetical year. The total volume of high-security entitlements is set at a fixed level for each month of the year. These month-by-month caps on high-security entitlements are set at levels that ensure that even during the driest of years, some water will remain in the ecosystem to protect ecological health. These monthly caps do not vary among years. In wetter years, additional water can be used in excess of the cap, as represented by low-security entitlements (the flex). As with high-security entitlements, the volumes of low-security entitlements are set at levels that will provide additional water for ecosystem support in wetter years, thereby providing the variability from year to year in water levels that are necessary to ensure ecosystem health.*

a scientifically prescribed, fluctuating regime of water flows in a river—which has proven to be very difficult to implement in South Africa—the Australian approach focuses instead on limiting how much water can be diverted and consumptively used from the river, leaving the remainder to the river ecosystem. A simplified version of this Australian approach is illustrated in figure 5-1.

I have a strong preference for this approach, which I will refer to as a "cap-and-flex" system for water allocation. By restricting how much water can be consumptively used, and not trying to continuously manage for a moving ecological target, this approach has proven to be considerably easier to implement. In addition to explicitly protecting freshwater ecosystems, the cap-and-flex approach can be effectively used to ensure that basic human needs are met by issuing high-security entitlements to those purposes, as discussed more fully under principle #3 below.

The limiting, or capping, of high-security entitlements provides certainty in how much water can be used in the driest years, while the flex feature enables water users to access additional water through low-security entitlements during wetter years, thereby fostering maximum economic productivity.

The cap-and-flex approach faces some of the same challenges discussed for the other two options, but its application in the Murray-Darling watershed of Australia suggests that it can be quite practical and robust. One major challenge in setting monthly cap levels will be to anticipate likely future changes in minimum water availability due to future climate changes. To protect freshwater ecosystems under all climatic conditions, care must be taken in setting the caps on the total volumes of high-security entitlements, and water managers should preserve the right to adjust those volumes as new scientific findings suggest.

Another big challenge is to know how much water can be allocated during flex years, when surplus water is available. In the Murray-Darling, large storage reservoirs in the upper areas of the system capture most of the natural water supply in the watershed, which comes primarily from snowmelt in the Great Dividing Range. These reservoirs have proven to be extremely important in measuring how much water is available each year (i.e., it is largely based on how much water is stored in the reservoirs) and in releasing water to downstream users at the time that they need it. Each spring, and continuing through the summer growing season, the volume of water to be allocated to low-security entitlements is adjusted largely on the basis of the volume of water stored in the reservoirs. The resulting low-security allocations will vary from month to month, or year to year.

The same cap-and-flex logic could be applied to lakes or aquifers, because they effectively store water much like a surface reservoir. It is relatively easy to measure the volume of water that has been added to an aquifer or lake from precipitation and runoff in recent months such that the volumes of water that can be allocated to low-security entitlements can be adjusted fairly readily.

If the water source is a river, and it is not possible to store large volumes of river flow, it will be necessary to develop other means for allocating water to low-security entitlements, such as by using winter snowpack levels to estimate the volume of water likely to be available during the forthcoming snowmelt period. Fortunately, advances in water monitoring and communications technologies are making real-time water allocation increasingly feasible. It may soon become commonplace to see variable water allocations being specified on a daily basis according to measured water levels in rivers, lakes, or aquifers, with water users accessing that information through the Internet.

Principle #3: Allocate a Specific Volume to Each User, Then Monitor and Enforce

In addition to setting limits on the total volume of water to be allocated as entitlements, it will be necessary to define the amount that each water user is authorized to use. In most countries, the right to use water—a water entitlement—is granted on a permanent basis, but with conditions specifying when the entitlement can be revoked or modified (see table 4-1 for examples of the conditions that are typically placed on a water entitlement). While these water allocation processes can and should be influenced by input from local community groups and water users—as discussed in chapter 4—a government entity or other public-service organization will usually need to oversee, manage, and enforce water entitlements. In other words, the regulation or governance of a water source needs to be managed as a public function at a community, watershed, state, provincial, or national level. Each individual water user applies to that public entity for the rights to use a specific volume of water and then complies with the rules and limits established by that public entity. The public entity—as the ultimate owner and arbiter of water—can revoke those entitlements when rules are violated. The entity might also modify entitlements when the values and priorities of the water user community change and a reallocation of water is desired.

When communities or countries begin instituting a water allocation or entitlement system for the first time, considerable attention will need to be given to any existing community-based laws or systems that have historically been used for water sharing.[4] Water agencies must be careful to avoid unnecessarily disrupting social and cultural systems and norms that might have been used effectively for hundreds or even thousands of years. One solution is to issue one water entitlement to an entire community, district, or other group of water users, allowing them to distribute the water among community members in their traditional manner. Most importantly, no one's right or access to water should be lost in the transition to a new allocation system simply because of inability to pay, illiteracy, religious or cultural beliefs, or other circumstances.

Quantifying Water Entitlements

When the cap-and-flex approach described above is applied, the volume of each individual high-security or low-security entitlement will need to be quantified. This quantification should be defined for each of the 12 months of the year, because water availability can vary substantially from month to month.

High-security entitlements are designed to ensure their fulfillment at all times. Some portion of the high-security entitlements should be reserved or allocated to the types of uses that the community of water users determines to be of greatest public value, thereby ensuring that these water needs will always be met. For example, basic human needs for water should be guaranteed by allocating a high-security entitlement to each and every community or individual. Other priorities for high-security allocations might include water that is needed for critically important public services, such as hospitals, firefighting agencies, schools, energy-generating facilities, and other socially valued services.

After these high-priority public needs are met, some portion of the available high-security water entitlements can be issued to commercial uses. Commercial water users that require a constant supply of water, such as manufacturers or farmers growing perennial crops, will likely want to secure high-security entitlements that can ensure a reliable supply of water.

Low-security entitlements provide an important supplement to the high-security allocations because they can enable water to be used to its maximum potential during wetter years. Unlike the high-security entitlements, for which a fixed volume of water is guaranteed in every month and year, low-security entitlements will provide their holders the full volume of their entitlements only during the wettest of years. In drier years, low-security entitlement holders will receive only a portion of their full entitlement.

A practical way of issuing low-security entitlements for each month of the year begins by determining the total volume of water expected to be available in the month during the wettest of all years, and then subtracting the volume of water expected during the driest year. This "surplus" monthly volume represents the *maximum* volume of water that could be allocated in each month for low-security entitlements. However, as with high-security entitlements, a decision needs to be made about how much of this monthly surplus will be allowed to remain in the freshwater ecosystem to support its ecological health, versus how much can be consumptively used by low-security entitlement holders. This additional and annually varying volume of water reserved for ecological purposes can be extremely important in sustaining fisheries and other benefits of healthy freshwater ecosystems, and it should be explicitly accounted for.

The resulting low-security allocations will vary from year to year, according to water availability. Typically, these changing allocation levels are communicated to water users as a percentage of the full volume of low-security entitlements; for example, during one year, an entitlement holder may receive 70 percent of his or her low-security entitlement, but in another year may receive only 20 percent or even less. Key to effective implementation of these low-security allocations will be the ability to forecast how much can be allocated in coming months, and communicating the allocation amounts to all low-security entitlement holders in a timely fashion.

Regardless of the approach used to allocate water entitlements, it will be essential to monitor water use and to punish violators of water entitlements if water management is to be successful. The Tarim River in far northwestern China provides a compelling illustration of this point. Following the end of the Cultural Revolution in China in the late 1970s and with the emergence of new agricultural policies, irrigated cotton farming began expanding rapidly in the watershed. The Tarim's water flows were fully depleted in nearly every year, causing serious ecological degradation and creating conflicts among cotton, wheat, and rice farmers.

The Xinjiang Tarim River Basin Management Bureau sets annual allocations of water in the watershed, but widespread illegal use of water has perpetuated conflicts, and the management bureau has found it very difficult to control violations. In August 2006, two officers from the management bureau tried to persuade farmers in the middle reaches of the river to remove their illegal irrigation canals. The officers were beaten and nearly drowned by the villagers.

Frequent water shortages, caused by the illegal use of water as well as highly inefficient irrigation practices that waste a great deal of water, have caused the Chinese government to consider some very expensive options for increasing the water supply in the Tarim watershed.[5] One of the proposed options involves pumping water from the Bohai Sea to an altitude of nearly 1,300 meters (4,300 feet), desalting it, and then piping it through numerous mountains to the Tarim watershed. In total, water would be transported over 5,400 kilometers (3,400 miles), farther than the distance from New York City to Los Angeles.

This Tarim story further accentuates the importance of local community engagement in water governance. When a community of water users does not understand or support the rules imposed on water allocation, it will be exceedingly difficult to gain cooperation or to police violators. Some degree of community peer pressure—along with responsible self-regulation—will be essential.

For hopeful examples, we can look to the acequia culture of irrigation management that has evolved over the past 10,000 years. It began in the Middle East and then was spread into southern Spain by the Moors, and then by the Spanish into the American Southwest. This communal system of water sharing and irrigating was a response to the scarcity of water in arid regions, and it has been key to the survival of many agricultural communities. In his book *Mayordomo*,[6] Stanley Crawford details the day-to-day interactions among a community of farmers in northern New Mexico who share the labor, cost, and responsibility of managing their shared acequia irrigation system. Community members join in the hard labor of maintaining earthen ditches, and the camaraderie formed through this ditch work has been very effective in discouraging water cheating within the community.

Principle #4: Invest in Water Conservation to Its Maximum Potential

Before pursuit of any of the water supply tools outlined in chapter 3, every possible effort should be made to first reduce water consumption. Every bucket of water saved through conservation or improved water-use efficiency is a bucket of water that does not have to be supplied or created with expensive infrastructure or technology.

In chapter 3 I described a variety of water-conserving activities that can be applied in cities, industries, or on farms. I also made the point that water conservation is by far the least expensive way of addressing a water shortage. Cost is only one argument for water conservation, however. By consuming less water, we can leave more in freshwater ecosystems, thereby sustaining healthy fisheries, recreational opportunities, and the many other benefits of healthy ecosystems. Investing in water conservation can also help to avoid the need to pursue other environmentally damaging water supply options such as overdepleting local water sources, building water importation pipelines that rob distant watersheds and aquifers of their waters, or having to burn fossil fuels to electrify water desalination plants.

Though you are not likely to hear the reasons expressed publicly, there are many reasons why governments and communities have not sufficiently invested in water conservation, including these:

- Water conservation is socially challenging because it requires changing the water-use behaviors of many individuals, as contrasted with a top-down decision to build a reservoir or pipeline that can be made unilaterally by a government agency.

- Most water agencies around the world are dominated by civil engineers, many of whom may be unfamiliar or uncomfortable with marshaling the social engineering of water conservation campaigns.
- Public and private water suppliers depend on water sales to support their annual budgets, so conserving water runs counter to their interest in making money.
- Many business leaders do not want to make a big deal about water conservation, for fear that it will send a message that their community is short on water, thereby discouraging new businesses from moving to the area.
- Politicians who build large water infrastructure projects such as dams can gain votes from a populace that does not understand that better options exist for addressing their water shortages.

These obstacles to water conservation are surmountable. But overcoming these obstacles will require that individuals, businesses, and community groups organize themselves and advocate forcefully for water-conserving solutions that are cost-effective, environmentally friendly, and sustainable for the long term.

Principle #5: Enable Trading of Water Entitlements

When water is in short supply, the ability to buy, lease, or barter for water within a community of water users can enable those who need more water to acquire it from those who may have some to spare. Informal water trading—in which neighbors exchange water through some form of bartering—has been taking place on farms and in villages around the world for thousands of years. More recently, water markets involving the buying, selling, and leasing of water entitlements have emerged in a number of countries. In many ways these modern water markets resemble financial stock exchanges; in a water market, the commodity being traded is the right to use water, that is, a water entitlement.

To envision how this works, consider, for example, a farmer who does not possess a water entitlement sufficient to irrigate a crop during an unusually hot and dry period, and therefore wants to lease additional water for a few months. Another farmer may want to expand his farm to grow more crops, and wants to acquire more permanent water entitlements for this purpose. If there are other water users in the area who do not need to use their full entitlements of water, they may decide to lease or sell all or some of their water entitlement to those who need more water. Through this process of water trading, those who value water more highly are able to acquire water from others who would like to receive monetary or other forms of compensation for their water entitlements.

Before further discussion of the merits of water trading, some important cautions need to be stated. Many concerns have been expressed about water markets, and they need to be given careful consideration. Most of these worries center on the possibility that water entitlements will be bought up by wealthy entities—water speculators, big corporations, or cities—leaving poorer people or freshwater ecosystems without water. This is a very real possibility if appropriate regulatory controls and other essential water governance functions are not put into place explicitly, and early. Every individual or family must be guaranteed an inalienable entitlement to enough water to meet basic needs, as discussed under principle #3 (water allocations) earlier. Those entitlements should not be tradable. Similarly, as discussed under principle #2 (capping total consumptive use of water), a sufficient volume and flow of water should be reserved or protected from trading to ensure ecological health. Water buyers should not be allowed to hoard water by acquiring water entitlements and not using them, and they should not be allowed to transport water out of a stressed watershed. Most importantly, nobody should ever be forced to sell their water entitlement, or have it taken from them unwillingly.

Water trading can also create undesirable impacts on agriculture if not properly anticipated and managed. One direct effect of transferring water out of agricultural use can be a reduced capacity for food production. Diversion of water away from agriculture can also reduce rural employment, which can be particularly undesirable in poor rural communities. If the farmers selling their water rights are members of a communal water supply system such as an irrigation district that maintains shared water infrastructure, the loss of too many irrigators due to water sales can place a heavy burden on the fewer irrigators that remain, as they must bear the ongoing maintenance costs for the infrastructure.

However, if these impacts are properly addressed and managed well, the ability to trade the rights to use water can be quite beneficial. In places like the Murray-Darling watershed of Australia, water markets have proven quite useful in enabling exchange of water entitlements among farmers, cities, and environmental interests, using both permanent sales and temporary leases. The benefits have been well documented. Farmers have been able to access additional water when they need it, such as at the end of an irrigation season, or gain a new source of income by selling or leasing their entitlements. By buying or leasing water from willing sellers, cities have been able to access additional water supplies in a way that is cost-effective and avoids having to further deplete local water sources. Environmental interests have been able to buy entitlements from water users and leave the water in freshwater ecosystems for ecological benefits.

Another major benefit of water markets has been their ability to stimulate water conservation, particularly when the holder of a water entitlement can sell or trade any saved water. For instance, when a management authority was formed in central Texas to regulate use of the Edwards Aquifer, farmers were given the opportunity to sell half of their water entitlements if they no longer needed the water, providing a huge stimulus for improving irrigation efficiency.

Setting limits or caps on total consumptive use of water and issuing a water entitlement to each user are essential prerequisites for high-performing water markets. Without a limit on total use, there is no imperative to share or trade, because users can take as much as they want or need. In the absence of quantified entitlements for water use, the trading parties will not know how much can be reliably exchanged.

In sum, the undesirable consequences of markets can usually be adequately addressed through governmental regulation and proper foresight, thereby enabling the benefits of water trading to be realized without harming poor people or ecosystems. However, appropriate controls and regulations, supported by strong governance systems, must be put into place before allowing trading to proceed.

Principle #6: If Too Much Water Is Being Consumptively Used, Subsidize Reductions in Consumption

I have said it a few times already, but I will say it again: it is much easier and less costly to constrain water consumption before it becomes excessive. Unfortunately, hundreds of water sources around the world are already being depleted far beyond a level that is safe or ecologically sustainable.

When severe water scarcity exists, one of the quickest ways to gain relief is for governments to step in and retire some portion of existing water entitlements to reduce the risk of water shortages. This can be accomplished through forced regulatory reductions in water use, but it is far more preferable to do this by compensating entitlement holders, that is, by buying water entitlements from willing sellers.

A government buyout does not require the existence of a water market, but a market can certainly help because water users will already be accustomed to the selling of water. In the absence of a market, the government will need to negotiate appropriate levels of compensation for those willing to use less water.

Reductions in consumptive use can also be subsidized through governmental investment in improved water-use efficiencies. Because agricultural irrigation is usually the biggest water consumer in water-stressed regions, particular attention should be given to opportunities to invest in improved irrigation practices and technologies.

Principle #7: Learn from Mistakes or Better Ideas, and Adjust as You Go

One of the great advantages of being human is our ability to learn. If we also possess a modicum of humility, we will have the ability to recognize and acknowledge when we are wrong, or not as right as we would like to be, and adjust our behaviors accordingly.

We should always assume that any water plan, water budget, or scientific assessment of ecological water needs will be at least partially wrong. We should also assume that our values and needs for water will change over time. These realities strongly suggest that we need to be ready for change or, more appropriately, always on the lookout for opportunities to do better.

Two strategies will be particularly important in facilitating the ability to learn and adjust over time. One is to enable water plans to be revisited on a regular basis. As discussed in the next chapter, the U.S. state of Texas revisits its regional and statewide water plans every 5 years. Another important strategy is to review individual water entitlements regularly. By reviewing water licenses at regular intervals, the government of South Africa has created a mechanism by which it can make adjustments in the volumes of entitlements over time.

Get Ready for Change

Many readers of this chapter will be quick to dismiss its principles as implausible, given the system of water management presently existing within their country or community. As one of my colleagues once remarked, "We don't all get to rewrite our constitutions like South Africa did after the fall of apartheid."

Governments do tend to get stuck in their ways, and political leaders are reluctant to advance changes that are too sweeping, or too fast, for fear that they will be run out of office. This is one of the more valuable benefits of water governance systems—they spread the responsibility for decision making, and they engage water users in that process.

History is full of telling examples of substantial shifts in governmental policies that have followed catalytic events or even gradual shifts in key social or economic conditions over time.[7] When the Cuyahoga River in the United States caught fire in 1969 because of a heavy concentration of oil and other chemicals in the water, it helped to set off a sea change in U.S. environmental legislation. When hundreds of kilometers of the Darling River in Australia became toxic in 1991 because of excessive nutrient pollution and depleted river flows—setting off a smelly and widespread fish kill—major water reforms including a cap on water entitlements were soon to follow. After decades of growing concerns over water pollution in Chinese rivers, the State Council made a strong commitment to pollution control in its 11th Five-Year Plan (2006–2010) and followed with intent to invest $60 billion on urban wastewater treatment in its most recent plan.

Who knows what water reforms may come to the Nile River in the aftermath of the social uprisings in Egypt? What changes may be in store for the Indus River if India and Pakistan decide to renegotiate their international treaty for that river?

Most successful water reforms have shared an important common element: active dialogue and debate had already been taking place in the years preceding the moment when changes could finally be put into place. In this sense, dissatisfaction or even conflict over water can be viewed as a useful precursor to change. The important lesson for water users or other citizens seeking change in their water governance is to be ready for it when the opportunity comes.

CHAPTER SIX

Bringing Power to the People

In April 2000, a 17-year-old student was shot dead by military police in the central plaza of Cochabamba, Bolivia. He had been protesting, in union with tens of thousands of other city residents, that his water cost too much.

Cochabamba, Bolivia's fourth-largest city, is nestled into a high Andean valley in the center of Bolivia. The valley is commonly referred to as "Bolivia's granary" because of its agricultural bounty of grains, potatoes, and coffee. Before 2000, Cochabamba was relatively unknown outside of Bolivia, but in that millennial year the city grabbed the world's attention. Cochabamba was embroiled in a water war. Rioters had seized the central plaza, blockaded incoming roads, and launched a worker's strike that temporarily shut down the city's economy.

The conflict was sparked by a sharp increase—averaging 35 percent or about $20 per month—in the price that city residents were paying for their water.[1] That rate hike—amounting to more than many poor families were paying each month for food—suddenly made drinking water unaffordable for many. As the water company threatened to shut off water to those who were not paying their bills, city residents took to the streets in protest.

The Cochabamba water war has been presented in news media and documentary films as a warning of what happens when large corporations are allowed to privatize water supplies.[2] In 1999, the Bolivian government contracted with a private consortium including the Bechtel Corporation and other companies[3] to administer the public water supply system of Cochabamba. The 40-year, $2.5 billion concession was intended "to provide water and sanitation services to the residents of Cochabamba, as well as generate electricity and irrigation for agriculture."[4] As part of the deal, the private consortium was required to absorb $30 million in debt that had accrued during the Bolivian government's inadequate management of the water system, make substantial improvements in the city's debilitated water distribution system, and increase water supplies by building a new storage reservoir. According to the consortium's representatives, the rate hike was instituted as the means for funding these projects.[5]

There have been many debates and critiques of things that went wrong in Cochabamba. Some see a silver lining in the Cochabamba story, because the people of Cochabamba were ultimately victorious in their protest and the Bolivian government has canceled the private consortium's contract. However, the people of Cochabamba still live under the dark cloud of a dilapidated water system, and half remain without piped water service.[6]

Many popular accounts of the Cochabamba story have been told in a way that strongly suggests that private corporations should be excluded entirely from providing a public service such as distributing water to city residents. Some activists have gone so far as to suggest that no corporation should be allowed to hold a water entitlement for fear that wealthy corporations could hoard water, to the detriment or exclusion of other citizens or the environment.

There are compelling counterarguments. Most people on our planet rely upon goods or services provided by the private sector that require water in their production or operation. Companies will not continue to produce those goods if they cannot be assured of water supply. Corporations can also play a constructive role by providing resources or expertise in situations where governments are weak and incapable of properly carrying out water functions.

I believe that it is very helpful to view these issues through the lens of water governance. Returning to my stone-balancing metaphor from chapter 4, we must in each instance seek the right balance among government, civil society, and the private sector. What conditions, agreements, contracts, and other constraints are needed to protect against misuse of water or inequitable use? How do we integrate civil society into the dialogue in a constructive manner?

Many water veterans assert that water governance has been pervasively failing because it has been too heavily controlled inside of governments, with too little external engagement. Historically, water planning, priority setting, and decision making have been too far removed from the local communities that are affected by such decisions. As so well illustrated by Cochabamba, these plans and decisions are too often made in capital cities by bureaucrats and technicians who have spent little or no time in the local communities that depend upon the water. When their edicts are not well understood or are perceived to be inappropriate to the local situation, it creates a feeling of disenfranchisement among those who are expected to behave according to plans and rules issued by their governments. The consequence—in both highly developed and developing regions of the world—is ubiquitous cheating by water users, inadequate reporting of use, conflict among competing water users, and poor or failed implementation of water plans and projects.

Bringing People to the Center

Elinor Ostrom, writing in her book *Governing the Commons* that helped her win the Nobel Prize in Economic Sciences, resisted the premise that "common pool resources" such as water are inevitably destined to fall victim to a tragedy of the commons. Ostrom advised that even in absence of governmental regulation, some communities have taken collective action to manage resources. Ostrom pointed to the communities of irrigators in the northern Philippines known as *zanjeras*. The zanjeras share the water of the Bacarra-Vintar River, using a communal system that assigns shares of water to each community member. One of the key obligations of each community member is to share in the labor of maintaining irrigation canals and a dam built across the river to divert water into the irrigation systems. During the rainy season each year, the river destroys the dam, which is constructed of bamboo poles, banana leaves, sand, and rock. Repairing the dam is arduous and rather dangerous work, involving several hundred people working in boat teams. That difficult, communal work has done much to create a bonding trust and camaraderie among the zanjeras. Even though they have faced many occasions of water shortage, when asked what water problems they face, none of the zanjeras expressed complaint about the way water was allocated or the fairness of water distribution.

But many did complain about the hardship of repeatedly repairing the dam!

Ostrom's primary conclusion was that *key management decisions should be made as close to the scene of events and the actors involved as possible.* This does not necessarily imply that national, state, or provincial governments cannot manage water well. But it does suggest that water governance in most places needs to be fundamentally restructured so that local communities and interested citizens play an integral role in water planning and allocation. The challenge of fitting a proper water governance frame to the management of local watersheds and aquifers will not be easy. Getting it "close to the scene of events and actors involved" will not always mean that conversations should only take place in local villages and towns. But two considerations will be paramount, both quite personal in nature. One measure of the success of water governance will be the ability of any citizen—farmer, industrialist, fisherman, or homeowner—to communicate his or her water-related values, concerns, and needs either directly or through a trusted representative participating in the water governance system. The other measure will be whether or not individuals embrace a sense of shared responsibility for the success of water governance, as reflected in compliance with water allocations and other rules and decisions.

Some really interesting experiments in local- or regional-scale water democracy are already under way. These experiments in water democracy have placed water users and other concerned citizens at the center of water governance. Below I describe two of these governance experiments in some detail. To be honest, none of the experiments in local water democracy is working perfectly. *But this is exactly how democracy is supposed to function.* It is supposed to be inclusive, transparent, experimental, and adaptive. That means messy, slow, and oftentimes inefficient. The magic of citizen-centered water governance is that it gets people talking to each other about water. They begin to examine how water is being used and who is using it, and it gets them thinking about ways to unlock some of the challenges or impediments to sustainable water management.

There are few experiences in life so daunting and empowering as realizing that our water future is now in our hands. Only in this manner can our water sources be managed with the caring and sharing necessary to avoid, or resolve, scarcity.

Texas: An Arranged Marriage of Water Stakeholders

A severe drought in the 1950s sounded an unwelcome wake-up call in the American state of Texas. Across the state, rainfall was 30 to 50 percent less than normal. Community water systems ran dry, causing city officials to send water truck convoys as far as Oklahoma to find water. Cowboys pushed their cattle north into Kansas to find green grass, but the drought eventually caught up with them there as well. Entire cattle herds died of thirst or hunger or were sold for slaughter, and crops withered in the field, causing thousands of farms and ranches to go broke. By the time the decade-long drought finally broke in 1957, more than half of the state's farmers had packed up and left.

In response, the state created a new Texas Water Development Board in 1957, charged with projecting water supply needs over a long horizon of 50 years, and provided funding to build new water projects. The first statewide water plan was published in 1961, and eight updates have followed. The state water plan serves as the primary guide for water development projects in Texas. According to plan, more than 180 major water reservoirs have been built in the state since 1957.

By the 1990s, the needs and values concerning water in Texas had shifted in some fundamental ways. Texas cities were among the fastest growing in the country, and their water needs grew with their populations. Urban water managers had begun questioning whether big regional reservoirs were always the right answer to their water needs, particularly given their high costs. Burgeoning urban populations were growing more concerned about the environmental impacts of water projects. There was a growing sense that the state water plan, fashioned in the capital city of Austin, was not sufficiently attuned to local needs and opportunities.

The Water Development Board tried to address these concerns by more actively engaging with the state's Parks and Wildlife Department and a number of stakeholder groups. The 1997 version of the state water plan was in fact labeled the "Consensus Based Water Plan." But something was still missing.

"It didn't really have buy-in from the local communities, from those that really know what their needs are and how best to meet those needs," said Cindy Loeffler of Texas Parks and Wildlife Department. "It still kind of missed the mark. . . . We thought it was great here in Austin, but you know, El Paso and Beaumont and Houston and Dallas, those cities had different ideas. To them, the water plan was just shelf art."

Coincidentally, another drought hit the state in 1996, just as the "consensus-based plan" was being printed. The drought came on hard and fast, leaving reservoirs depleted and communities across the state running out of water. "That drought got a lot of people worried, and they started asking, What do we need to do?" remembers Robert Mace of the Water Development Board. "And then a political leader said, well, we've got the state water plan, let's open it up and take a look at it. I think a lot of people were surprised when they saw what was in there. A lot of the communities said, geez, nobody asked us about this—and that's not what we need to do! There was clearly a disconnect between the water plan and the people that needed to implement it."

The stage was set for sweeping change in the way the state plans for its water needs.

Opening the Doors of Water Planning

Texas Senate Bill 1, passed in 1997, created sixteen regional planning groups across the state. These planning groups are to include "representation from the interests comprising that region, including but not limited to the public, counties, municipalities, industries, agricultural interests, environmental interests, small businesses, electric generating utilities, river authorities, water districts, and water utilities." The planning groups are composed of about twenty members each, a political marriage arranged through appointments made by the Water Development Board. According to Senate Bill 1, they are charged with the task of preparing 50-year water plans "that provide for the orderly development, management, and conservation of water resources and preparation for and response to drought conditions in order that sufficient water will be available at a reasonable cost to ensure public health, safety, and welfare; further economic development; and protect the agricultural and natural resources of that particular region."

Here is another way of describing the task of the regional planning groups: they are to develop—for each of the water sources within their planning jurisdictions—projections between water deposits and expenditures over the next 50 years and recommend an optimal mix of strategies to fix any gaps in those water budgets. As Robert Mace puts it, "The state provides the guard rails, such as by projecting population growth or running the groundwater models, but in terms of identifying what needs to happen, that moved from our agency down to the regional planning groups."

The sixteen planning groups went to work immediately after appointment. The members of these groups soon learned that it was not going to be easy to build a shared vision for their community's water future. Carolyn Brittin, who for more than a decade shepherded the regional planning process on behalf of the Water Development Board, remembers that "when those stakeholders sat down together, they quickly realized that they weren't going to be able to get everything they wanted. They learned that they were going to have to work it out with everyone else in the room." Robert Mace concurs: "Some relationships start to form. They gain some understanding of the person across the table. Okay, the world's not going to end up my way, and it's also not going to end up your way, but is there a third way that will make us all happy?"

With the memory of recent droughts spurring them forward, the regional planning groups completed their first 50-year water plans on schedule. The first integration of all sixteen planning regions in the 2002 State Water Plan proposed investments worth $23 billion that were designed to meet projected water shortfalls to 2050. Each of the sixteen regional plans included a diverse mix of water strategies to balance local water budgets, ranging from building new reservoirs or water importation pipelines to water recycling. Some plans called for clearing vegetation such as juniper shrubs from watersheds to lessen the volume of water lost to these thirsty plants, which have become more abundant in recent decades.

Notably, the 2002 State Water Plan—the first product of the regional stakeholder process set into place by Senate Bill 1—called for a sizable portion of the water deficits in the state to be reduced with water conservation. Instead of relying entirely on measures to increase water supply, the stakeholder-scripted plans called for 14 percent of the state's water deficit to be met by reducing water consumption.

This emphasis on demand-side water management grew over the next two water planning cycles, amounting to 23 percent of the water solution in the 2007 plan, and 24 percent in the 2012 plan. This shift in strategy was in large part stimulated by growing attention to cost-effectiveness. As the stakeholder groups began realizing that urban and agricultural water conservation could close their water gaps at a third of the cost of building a new reservoir and a tenth of the cost of building water import pipelines or desalination plants, they started placing heavier bets on water conservation as the best way to solve their water challenges.

This gravitation toward saving water is also a reflection of a growing conservation ethic in the state. Many Texans simply do not want their rivers to be dried up any further, and they are willing to be more conservative in their water use if it helps avoid taking more water from rivers. One of the great benefits of citizen-centered planning and governance is that it can move swiftly in the direction of changing values, oftentimes much faster than governments are willing to respond.

By implementing a regional planning process, the state of Texas has been able to implement some of the important water sustainability principles discussed in the previous chapter (see table 6-1 for my own ratings of the state's performance on these principles). The technical and legal details of water allocation have been well managed and monitored by the state's water agencies for many years. The stakeholder-based planning process has helped water users and other citizens understand how that allocation process aligns (or doesn't align) with their vision for the future. By revisiting their plans every 5 years, the regional groups are able to integrate new information, new priorities, and new opportunities, adapting as they go. And while potential exists to do more with water conservation, the fact that nearly a quarter of the water deficit statewide is to be shrunk by reducing water consumption is quite notable.

One principle that has received inadequate attention among these stakeholder groups and the Texas government to date is the need to place a limit on total water consumption when water sources are being heavily depleted. Water use statewide is projected to increase by more than 20 percent in coming decades, which will place more strain on many water sources that are already nearing exhaustion. Recurring high levels of water depletion—like those shown in figure 1-4 for the Brazos River—are placing the state's economy and the health of its rivers in jeopardy. As discussed in chapter 5, setting limits on consumptive use can be politically

Table 6-1 Sustainability Scorecard for Texas

Sustainability Principles	Little to No Progress	Notable Progress	Strong Performance
Principle #1: Build a shared vision for your community's water future.			√
Principle #2: Set limits on total consumptive use of water.	√		
Principle #3: Allocate a specific volume to each user, then monitor and enforce.			√
Principle #4: Invest in water conservation to its maximum potential.		√	
Principle #5: Enable trading of water entitlements.		√	
Principle #6: If too much water is being consumptively used, subsidize reductions in consumption.	√		
Principle #7: Learn from mistakes or better ideas, and adjust as you go.			√

contentious, and it can be challenging to quantify how much water needs to be left in a river to protect its ecological health. However, the ecological and economic costs of overallocation can be very disruptive and damaging. The water crisis unfolding in the Guadalupe–San Antonio watershed of Texas highlights the importance of setting limits on total water consumption, an issue that will surely continue to confront the state's water agencies and regional water planning groups in coming years.

A Court-Mandated Moratorium on Water Allocation

The U.S. Endangered Species Act (ESA), established in 1973, is designed to protect critically imperiled species of plants and animals from going extinct due to "economic growth and development untempered by adequate concern and conservation." The ESA is an expression of the intent of the American people to protect the country's rich natural heritage, which is of "esthetic, ecological, educational, recreational, and scientific value to our Nation and its people." Since the ESA's creation, it has proven to be one of the country's most effective laws for protecting the ecological health of freshwater ecosystems.

In March 2013, a federal judge ruled in a lawsuit that the state of Texas had violated the U.S. Endangered Species Act by overallocating water in the Guadalupe and San Antonio Rivers. These rivers provide freshwater inflows into San Antonio Bay, the winter home of highly endangered whooping cranes. During a drought in 2008–2009, very little freshwater reached the bay, resulting in greatly increased water salinity and lower abundance of the blue crabs and wolfberries that cranes feed upon. At least twenty-three cranes died from malnourishment. The federal court placed an injunction against the issuance of any more water entitlements in the Guadalupe and San Antonio Rivers.

In its 2011 regional water plan,[7] the South Central Texas Regional Water Planning Group, which is the stakeholder group responsible for water planning in the Guadalupe and San Antonio watersheds, noted, "A key concern in the South Central Texas Region is that of threatened and endangered species." The plan also included the stakeholder group's responses to a number of public comments received on their 2010 draft. One commentator suggested, "The [plan] does not adequately address the needs of whooping cranes." The planning group responded: "The actual needs of whooping cranes are not known in sufficient specificity."

This situation may improve considerably in the next 5-year iteration of the regional plans, however. In 2007, in response to gaps in understanding of environmental water needs across the state's rivers, the state legislature in Texas passed Senate Bill 3, calling for the appointment of science and stakeholder committees to make recommendations on the environmental flows needed to maintain the ecological integrity of the state's rivers. Similar to Senate Bill 1 that created regional water planning groups, Senate Bill 3 also clearly expresses the intent to foster collaboration and seek consensus among stakeholders. The legislation calls for scientific committees to be appointed for each river. These committees are to prepare an assessment of environmental water needs, which stakeholder committees will then review before making a recommendation to the state agency responsible for water allocation. These recommendations are intended to balance the needs of the environment with the needs of water users.

Unfortunately, finalization of the environmental flow recommendations for the Guadalupe and San Antonio Rivers did not come until September 2011, a year after the water plan was completed by the South

Central Texas water planning group. As a result, those environmental flow recommendations were not integrated into the water plan, but they will be incorporated into the next 5-year edition of the regional water plan. "Those environmental requirements have now been adopted by the state, and we're going to use them in our current planning," says Con Mims, who chairs the regional planning group. "We'll crank those criteria into it."

In the meantime, though, the fate of the regional water plan is highly uncertain. Some of the biggest solutions called for in that plan would further deplete the Guadalupe and San Antonio Rivers by utilizing existing water entitlements more fully or relying on additional allocations from the rivers, which could further reduce freshwater inflows to crane habitat in San Antonio Bay.

Rebalancing the Water Budget of the Guadalupe and San Antonio Rivers

In its planning process, the South Central Texas stakeholder group had a lot more than whooping cranes to worry about. The region's population is expected to more than double in the next 50 years, and water use for municipal and industrial purposes is expected to grow by nearly 60 percent. "The region has a huge variety of water demands, from agricultural to municipal to industrial to environmental," explains Mims. "The region's population is exploding, and new water supplies are dwindling. Those two things are causing a lot more people to want to sit in on the planning group." The group has increased from twenty-one members to twenty-nine already. "But at some point, you just have to shut the door and quit adding more members. At the same time, we want to make sure that all important interests are sitting at the table," says Mims. A 5-year term for planning group membership and regular turnover helps to keep the regional planning groups adaptable and responsive to changing needs and interests.

Some environmental interests in Texas have asserted that the regional water planning groups are stacked with stakeholders more interested in water development than environmental protection, asserting that there is imbalance in water governance.[8] In that light, the federal lawsuit over whooping cranes could be viewed as a civil society effort to rebalance power among the water planning stakeholders for the Guadalupe–San

Antonio watershed. This tension illustrates a key point about local, citizen-centered water governance: it will always need to be respectful of the values and laws held by the broader society, such as the desire to protect endangered species. But Mims is concerned that the lawsuit could seriously disrupt a consensus-based plan that has been very difficult to construct. "If that court ruling holds [it is currently under appeal] and the state has to give additional priority to freshwater inflows into San Antonio Bay for the sake of the whooping cranes, we would have to completely revamp our regional water plan. It would turn the plan on its ear."

In its most recent planning cycle, the regional group compared projected water needs with available water supplies and found a shortfall of more than 500 million cubic meters by 2060, meaning that the planning group needed to find 50 percent more water than is currently available. To make matters worse, scientific analysis associated with the federal court case has identified a possible need to reserve at least 185 million more cubic meters for freshwater inflows to the bay.[9] The bottom line: the currently available water is already overallocated, with many more demands coming.

The regional group's plan calls for a broad array of measures to address the gap between available supplies and projected needs. In light of the federal court decision, the planning group will likely need to take a hard look at some of their proposed solutions. They have already targeted water conservation measures to reduce their deficit by 15 percent, but there may well be room for more. Mims believes that desalination will be critical, particularly if other options called for in their plan are nullified by the court ruling. As discussed in chapter 3, one of the strong benefits of desalination is that it creates a new source of freshwater without having to further deplete existing water sources such as rivers.

Table 6-1 highlights two other principles that might be pursued in the Guadalupe and San Antonio watersheds. The Texans could explore ways to buy down the current level of consumptive use of water (principle #6). Doing so through regulatory action is highly unlikely, because the state's citizens are not very receptive to governmental regulation, particularly if it threatens to reduce existing rights such as water entitlements. They might instead establish a formal market to facilitate water trading (principle #5). This would enable the government or environmental interests to more easily purchase water entitlements from willing sellers, for dedication to environmental flows in the rivers, thereby reducing overall water consumption and providing relief to the bay ecosystem and the cranes. In fact, Senate Bill 3 directed the state to do exactly this in situations such as those in the Guadalupe and San Antonio watersheds: "In those basins in which the unappropriated water . . . is not sufficient to satisfy the environmental flow standards, a variety of market approaches, both public and private, for filling the gap must be explored and pursued."

The process of building consensus around specific actions to balance the water budget has caused a lot of strain within the regional group. "The planning group members work very, very well together," says Mims. "Where the problems arise is in the very small handful of projects that are controversial. Even one project can create all sorts of turmoil." Proposals for new reservoirs are particularly controversial in this region because of concerns over environmental impacts and high costs.

Fortunately, the South Central Texas planning group has already spent many long hours together, and they very well understand the challenges ahead. Their trust and respect for each other will be put to the test as the group grapples with these daunting challenges during their next cycle of planning.

Kenya: Strengthening Governance through Public-Private Partnerships

On the eve of Valentine's Day[10] in 2006, romantics across Europe were confronted with a taint on the roses they planned to give their sweethearts.[11] The Reuters News Service had just reported widespread social and environmental abuses in the cut-flower industry of Kenya, where the majority of flowers are grown for export to Europe. News stories told of workers underpaid for long hours, widespread sexual abuse, and health problems and environmental contamination resulting from pesticides and other chemicals used in the industry.

As social and environmental activists began to tune in to these reports, Lake Naivasha in Kenya quickly became a center of international attention and concern. Since the 1980s, commercial flower farms in Kenya have become the largest suppliers of flowers to the European market, shipping more than 96,000 tons of cut flowers a year, worth some $463 million.[12]

Lake Naivasha is a hub of this flower production, with more than thirty large farms lining its lakeshore, providing employment for thousands of local people.

The lake is globally renowned for its beauty and biodiversity. Situated in the Great Rift Valley, the lake receives most of its water inflow from the Malewa, Turasha, and Gilgil Rivers. The lake was recognized in 1995 by the Ramsar Convention as a wetland of international significance, and it has long attracted bird-watchers and other nature lovers to its shores. But in recent years, growing concern for the lake's health and a declining lake level has been voiced by many individuals living and working in the area. David Harper, a professor at the University of Leicester with long research experience in the area, told *The Times* in a 2006 interview, "Almost everybody in Europe who has eaten Kenyan beans or Kenyan strawberries, and gazed at Kenyan roses, has bought Naivasha water. The unsustainable extraction of water for agriculture, horticulture, urban and residential water supplies is sucking the lake dry."[13] Isaac Oumo Oloo, a local activist and safari guide, expressed similar sentiments: "We take this water to grow flowers and then ship them 5,000 miles to Europe so that people can say 'I love you, darling' and then throw them away three days later. To me that is an immoral act."[14]

While many people had expressed concern that lake level declines and river drying were being caused by overconsumption of water, the extraction of water from the lake and its tributary rivers had not been regulated until recently. Concerned with their water supplies and fearing damage to their corporate reputations, a consortium of flower companies began collaborating as the Lake Naivasha Growers Group in 1997. "A number of concerned growers recognized that unless there was some regulation within the industry, there was a risk of pollution with a very negative impact on the environment," remembers Richard Fox, the sustainability director for Finlay's Horticulture in Kenya. The grower's group soon expanded their focus into water allocation, and they opened their planning efforts to a much fuller array of water users and citizens that included government staff, tourism operators, small-scale farmers, pastoralists, fishing groups, public water suppliers, geothermal energy developers, researchers, and others.[15] The expanded group completed their first water allocation plan in 2005.

Their timing could not have been better. When Kenya rewrote its national Water Act in 2002, it gave strong emphasis to the formation of local "water resource user associations" (WRUAs). These citizen-centered water groups are overseen and supported in Kenya by the national Water Resource Management Authority (WRMA). WRMA has now officially recognized the Lake Naivasha Growers Group, along with other interested citizens, as the Lake Naivasha WRUA. The group's water allocation plan was formally adopted by the Kenyan government in 2010.

The 2002 Water Act in Kenya had instituted a number of other important reforms to aid water governance. The act strengthened the country's water allocation system, which is based on the issuance of water permits, and adopted the concept of creating a reserve of water in each watershed that cannot be allocated, to ensure that everyone in the country has access to water for basic needs and that enough water will remain in the country's freshwater ecosystems to sustain their health. The act also established a water user fee system, in which water users are charged according to their water use, for the express purposes of (1) enabling WRMA to improve overall water-use efficiency in the country, (2) collecting data on water availability and use, and (3) supporting water management activities including those undertaken by the WRUAs. Also of note is the way that water entitlements are to be monitored and enforced. Each WRUA is to provide staff for monitoring compliance with water permits and to collect water-use fees.

Lake Naivasha Water Allocation Plan

As a result of its deliberations over a water allocation plan for the Lake Naivasha watershed, the WRUA put into place a number of very important goals and regulations. Their shared vision for the watershed has been expressed as follows: "The goal of the water allocation plan is to establish a reasonable and practical framework for water allocation and water abstraction within the Naivasha Basin, which has been agreed and adopted by stakeholders, which is anchored in current policy and legislation and which aims to safeguard the natural ecosystems from over abstraction while supporting multiple demands on the water resources."[16]

The Lake Naivasha water allocation plan also set a goal of reducing total water use by 10 percent by December 2012. To facilitate attainment of this goal, the plan set up a permitting system for all water users and requires them to install water meters to measure their water use and report their use to WRMA.

The plan also set up a framework for reducing water allocations during times of water scarcity. Table 6-2 details the rules set up for those withdrawing water directly from the lake; similar rules are included for river and groundwater users.

Success in the Lake Naivasha watershed will largely depend on how the new water permitting process is carried out, however. The plan notes that some components of the water budget are not well understood and quantified, which makes sustainable water permitting very difficult. Very little governmental funding for water monitoring has been available, and the absence of an accurate water budget has frustrated the water permitting process. Without a fully developed water permitting system, the government is also losing an important source of revenue that could be used to address these shortcomings.

Table 6-2 Water-Use Restrictions for Lake Naivasha during Water-Scarcity Periods

Color Coding for Restrictions on Water Withdrawals	Lake Elevations (meters above sea level) and Associated Water Restrictions
Green = satisfactory Withdrawals allowed up to permit limits	Lake elevation higher than 1,885.3
Amber = stress Slight restrictions imposed	Lake elevation between 1,884.6 and 1,885.3 Domestic and public water supplies 100% Other uses 75%
Red = scarcity Severe restrictions imposed	Lake elevation between 1,882.5 and 1,884.6 Domestic and public water supplies 75% Other uses 50%
Black = reserve protection Withdrawals limited to basic human and livestock needs	Lake elevation lower than 1,882.5 Domestic and public water supplies limited to basic needs only (i.e., 25 liters per person/livestock unit per day)

Adapted from Kenya Water Resources Management Authority, *Water Allocation Plan—Naivasha Basin 2010–2012* (Nairobi: WRMA, 2010).

"The big challenge was that the whole water allocation plan rests with a government organization called the Water Resources Management Authority," says Fox, "and they are grossly understaffed and under-resourced, so that when we did a survey in 2010, we found that only about 4 percent of water users had a valid permit. This unregulated abstraction began to really make people realize that here you've got a very valuable industry sitting on a very fragile ecosystem with people's livelihoods depending on it, and we needed a plan for how this could be managed, because from an industry perspective there's significant business risk, both reputational and financial, if things aren't managed properly."

The response of private business interests to the water problems of Lake Naivasha exemplifies a rapid shift in corporate thinking globally about business risks and responsibilities related to water. Stuart Orr leads the corporate water stewardship program at WWF, an international conservation organization also known as World Wildlife Fund. Stuart has worked with some of the companies around Lake Naivasha and many others around the world, encouraging them to take greater responsibility for sustainable water management. He points out that water presents a unique challenge to business because it represents both a dependency and a risk. "It's a dependency because all business needs water in good quality and at the right time, and a risk because it is becoming more stressed, with more competing uses and continued weak governance. This corporate link to water makes it paramount for companies to work with others in the local communities to resolve their mutual water challenges—yet they are only beginning to wake up to this necessity."

Recognizing the vulnerability of Lake Naivasha's economy and ecological health, the prime minister of Kenya met with the Prince of Wales in 2009 to ask for help. The Kenyan and Dutch governments and some large British supermarket retailers subsequently provided funding to support private consultants and planning for the advancement of sustainable development in the Lake Naivasha watershed, which became known as the Imarisha Project (*imarisha* is a Swahili word meaning "to be reborn" or "to repair"). "We were able to bring this really strong private sector engagement in water management into an environment where everyone could contribute to a plan for managing the resources of Lake Naivasha sustainably," says Fox. The Imarisha Project is now attempting to create a long-term private funding source. "We see a contribution from every flower sold from Naivasha, from every kilowatt of energy that is being generated out of Naivasha, from the number of bed nights from tourism in Naivasha, and from fees on water use from the lake. These are sources of revenue we can tap, provided we get people to realize that everyone's interest is in sustainable management of Naivasha's resources." These private funds are expected to greatly enhance the WRUAs' ability to implement the water allocation plan more fully, particularly the collection of data needed to better quantify the water budget.

Based on these developments to date, I have scored the Lake Naivasha sustainability principles in table 6-3. Most of the less-than-perfect scores are due to the very early state of planning and implementation in this watershed. Prior to the adoption of the 2010 plan, there were no controls on water use. When viewed in that light, great progress has been made.

The situation in Lake Naivasha helps to illustrate the fact that some of the sustainability principles I have offered may not be relevant in all places. For instance, while the Lake Naivasha water allocation plan notes that water savings of 25 to 50 percent should be readily attainable on irrigated farms (Finlay's itself has reduced its use by 40 percent in recent years), it does not mandate such water conservation in any direct way. Instead, it is assumed that if water allocations are issued and capped in an appropriate and proactive manner, further water conservation by existing users may not be necessary, and the government would not need to subsidize reductions in water consumption. This situation very well illustrates the point that when water plans and allocation systems are implemented before a water source becomes overallocated, the governance opportunities are much less costly and much more politically palatable.

Looking to the future, the Lake Naivasha WRUA might anticipate growing interest in water trading. Once the total volume of consumptive use of water is capped for the watershed, any new users or existing users wanting to expand their use will be constrained by the cap. This could create a fertile environment for water conservation and trading. If, for example, water users are allowed to sell any water saved through conservation to other users needing more water, incentives for water saving will become available. From the experiences of other water markets,

Table 6-3 Sustainability Scorecard for Lake Naivasha, Kenya

Sustainability Principles	Little to No Progress	Notable Progress	Strong Performance
Principle #1: Build a shared vision for your community's water future.			√
Principle #2: Set limits on total consumptive use of water.		√	
Principle #3: Allocate a specific volume to each user, then monitor and enforce.		√	
Principle #4: Invest in water conservation to its maximum potential.	√		
Principle #5: Enable trading of water entitlements.	√		
Principle #6: If too much water is being consumptively used, subsidize reductions in consumption.		√	
Principle #7: Learn from mistakes or better ideas, and adjust as you go.		√	

we can expect that such trading would likely result in increasing levels of economic productivity, which in turn can provide new jobs and increased incomes. However, great care must be taken to ensure that subsistence uses of water, such as for drinking water and family gardens, are well protected, and that water trading does not impinge upon the reserve of water for basic human needs and ecosystem support.

When I asked Richard Fox what advice he would offer to other groups becoming involved in water governance, he pointed out that attention and enthusiasm can wane during wetter times. "The problem is, when the water levels are high, it flows completely out of people's minds that there ever was a problem. When drought hits again, there is great potential for water conflicts to arise. It's critically important to get stakeholders talking to each other. It's important for everyone to realize that they are one among a large matrix of people that depend upon water. Unless they individually feel that they're getting their fair share, or their voice is being heard, then the whole principle doesn't work. No one group, no matter how big or small, can achieve their objectives unless everyone else achieves their objectives from the process."

With Power Comes Responsibility

Texas and Kenya are but two of many places in the world where citizens and businesses are gaining stronger opportunities to influence their water future. These newly empowered water ambassadors are quickly learning that nothing is easy when it comes to water.

None of the newly forming local planning groups or water user associations will have the opportunity to start with a truly clean slate. They will all inherit long legacies of water use, and in many cases they will be handed water accounts that have been overspent historically. They also will not likely be given a great deal of opportunity—at least not right away—to make big changes in their government's policies for water allocation.

In this sense I am reminded of the immense challenges that Nelson Mandela faced when he became South Africa's new president in 1994. Mandela knew that if his citizens focused exclusively on the country's immediate problems, they might quickly become overwhelmed and pessimistic. Mandela instead inspired his countrymen to look out over a longer horizon, and to help him build a future in which they all could prosper.

That wisdom is especially pertinent to our efforts to resolve water scarcity. Many emerging water governance groups will be confronted with a woeful lack of information and will need to work with their governments to assemble a reasonably accurate water budget, similar to what has occurred at Lake Naivasha. It may take time to acquire sufficient and reliable information about the full range of options available for increasing water supplies or reducing demands on a par with the toolbox of options assembled for the Colorado River basin that was discussed in chapter 3. Pursuing implementation of the sustainability principles highlighted in chapter 5 will often require extraordinary levels of cooperation and coordination among government agencies, private businesses, and citizens with widely divergent understandings of water science, law, and policy.

Resolving water scarcity should not be viewed only as a problem-solving exercise, as a one-time task. Successful water governance is a process, a road that never ends. When performed well, it involves active learning, advocating, disagreeing, compromising, deciding, and adapting. If a citizen-centered water democracy is framed with that perpetual process in mind, and if its participants commit to staying at the table when issues become thorny or uncomfortable, the prospects for success are very promising.

CHAPTER SEVEN

How to Survive a Water Crisis: Murray-Darling Basin, Australia

Many of the Aboriginal peoples of Australia believe that all life emerges from pools of freshwater. It is not surprising that in a land of such extreme aridity and variability in rainfall, water would take on spiritual importance. The original Australians have lived with fickle and harsh climatic extremes for tens of thousands of years, learning to adapt and passing on this wisdom through generations.[1] Their understanding of weather fluctuations among seasons and years is reflected in their calendars, their language, their art, and their nomadic migrations. They have long known where to go to find edible plants or fish when the rains did not come.

The livelihoods of modern farmers in the region, by contrast, are tethered to irrigation canals. And when the Millennium Drought of 1997 to 2009 left those canals bone dry, they found themselves in grave peril.

During "the Big Dry," as the drought also came to be known, Australian farmers experienced a parching unlike anything they had ever seen or heard of. The drought was so severe that many of the rivers and streams in the Murray-Darling watershed (fig. 7-1) stopped flowing. Many farmers were left with no water for irrigation.

Some farmers sold off their entire herds of dairy cows as the animals became too expensive to feed with imported hay, while others watched mature fruit trees turn into firewood. Without a crop to sell, many began falling far behind in their debt payments to the bank. Lots of debt-ridden farmers had nothing of any value left to sell except their water entitlements. Once the water was sold and gone, only despair remained.

"People were starting to commit suicide," says John Conallin, a farmer's son, upon returning home to the Murray-Darling after years away at college studying environmental science and natural resource management. "Some of them felt such a sense of hopelessness that they saw suicide as their only way out. Even my dad, who had farmed for 50 years, was starting to wonder why he bothered to get out of bed every day."

Farmers in Australia have seen plenty of dry years. But by 2009, after more than a decade of severe drought, they were at their wits' end.

When I met Conallin in 2013, the drought had broken, but he was still deeply pained by the memories of the hardships his family and neighbors went through in the farming community of Deniliquin, which straddles the Edward River, a tributary of the Murray River (fig. 7-2). Conallin had been away for university studies for 14 years, missing the drought almost entirely except on short visits back home. After earning his undergraduate degree in Australia, he went on to Denmark to complete his master's and doctoral degrees. He said that by the time he finally returned home to the Murray-Darling watershed in 2009, "much of the river system had become disconnected and dried up into a series of stagnant pools. Fish were dying everywhere. We were trucking water into our farm. It was starting to look like it was coming to the end of it all."

The story of that tragic drought, and the way in which the Australians dealt with it, is one of the most intriguing and enlightening water histories that I know. As with most stories of water shortages, the water crisis took a long time to build in the Murray-Darling watershed, but the breadth of reform and adaptation that took place during the Millennium Drought was nothing short of remarkable. In this story are many lessons for other water-stressed communities around the world.

Living with Uncertainty

Australia has one of the most highly variable climates in the world. There are years when the Murray-Darling watershed receives more than 800 millimeters (30 inches) of rainfall, but some years see less than 250 millimeters (10 inches). This uncertainty of water supply has been a source of recurring distress for the region's farmers. Even after nearly two centuries of settlement, today's descendants of European colonists are still coming to grips with the unreliable climate of their new home. As Daniel

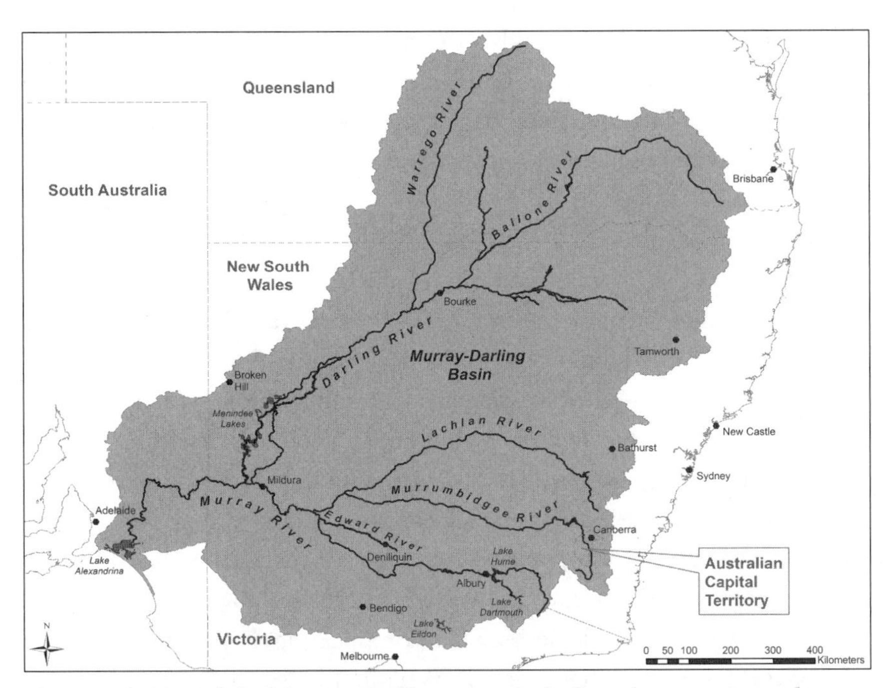

Figure 7-1. *Map of the Murray-Darling watershed of southeastern Australia. (Based on data from Global Runoff Data Centre of Germany and the Commonwealth of Australia, Department of Sustainability, Environment, Water, Population and Communities.)*

Connell writes in his book *Water Politics in the Murray-Darling Basin*, "At the beginning of the 21st century Australians are still struggling to define their relationship to the continent's climate and landscape."[2]

European settlement in the region began in 1836 with establishment of the colony of South Australia on the Adelaide Plains, not far from the mouth of the Murray River. Within a couple of decades, paddle-wheeled steamboats were pushing upstream more than 2,000 kilometers (1,200 miles) into the upper reaches of the Murray and Darling Rivers to haul wool, timber, and other resources to Adelaide's port (fig. 7-3). The discovery of gold in the 1850s brought thousands of new settlers into the watershed, along with a rapidly growing appetite for agricultural goods. But unpredictable rains proved to be the bane of Australian farming from the very beginning.

When farmers first began settling in the Murray-Darling watershed in the 1870s, they ignored the warnings of Surveyor General George

Woodroffe Goyder that droughts of the sort seen just a decade earlier would surely return. It did not take long for Goyder's predictions to come true, leaving the farmers' fields barren and forcing them to retreat downriver to Adelaide in search of another way to make a living.

The Rise of Irrigated Agriculture

The presence of a flowing river and plentiful sunshine was too alluring for many colonist farmers to resist for long, however. They quickly came to appreciate that irrigation was going to be essential to their success. Of the 460 mm (18 inches) of average annual rainfall on the watershed, only about 6 percent—equivalent to 32 billion cubic meters (BCM) or 26 million acre-feet (MAF)—ends up in the river system, because the severe heat and aridity evaporate water so quickly from the landscape.[3] The volume of river flow can vary enormously from year to year: 118 BCM (96 MAF) of water rushed downstream in 1956, but in 2006 only a paltry 7 BCM (5.7 MAF) of water was available. Further complicating irrigation in this part of the world is the fact that much of water in the river system comes from late winter snowmelt in the Great Dividing Range to the east, and it flows

***Figure 7-2.** Dead Murray cod (*Maccullochella peelii*) being hauled out of a dried-up stream near Deniliquin during the Millennium Drought. (Photo by John Lolicato.)*

***Figure 7-3.** The Murray River near Renmark, Australia. Paddle-wheeled steamboats have been traveling this stretch of river since the 1860s, originally for the shipping trade, but today they carry only tourists.*

away quickly in spring, well before it is needed for irrigation in summer and autumn. To overcome their water limitations, the farmers of the late 1800s built hundreds of small dams to capture river water for use during the summer growing season.

Much larger reservoirs were built in the twentieth century (fig. 7-4), providing greater certainty and stability in agricultural production. By the 1980s, 3 full years' worth of river flow could be stored in reservoirs, supplying water to an extensive network of farms and rural towns across the watershed. The Murray-Darling—hailed as Australia's food basket—today accounts for nearly two-thirds of all irrigated farmland in Australia, generating nearly half of all farm revenue in the country. Virtually all (96 percent) of the consumptive use of Murray-Darling water goes to irrigating cotton, rice, cereals, alfalfa, fruits and nuts, wine grapes, beef and dairy cattle, and sheep (table 7-1).[4]

Setting Rules for Water Allocation

As the practice of irrigation spread rapidly across South Australia, New South Wales, and Victoria, the states quickly recognized that rules for water sharing—both among the states and among water users within

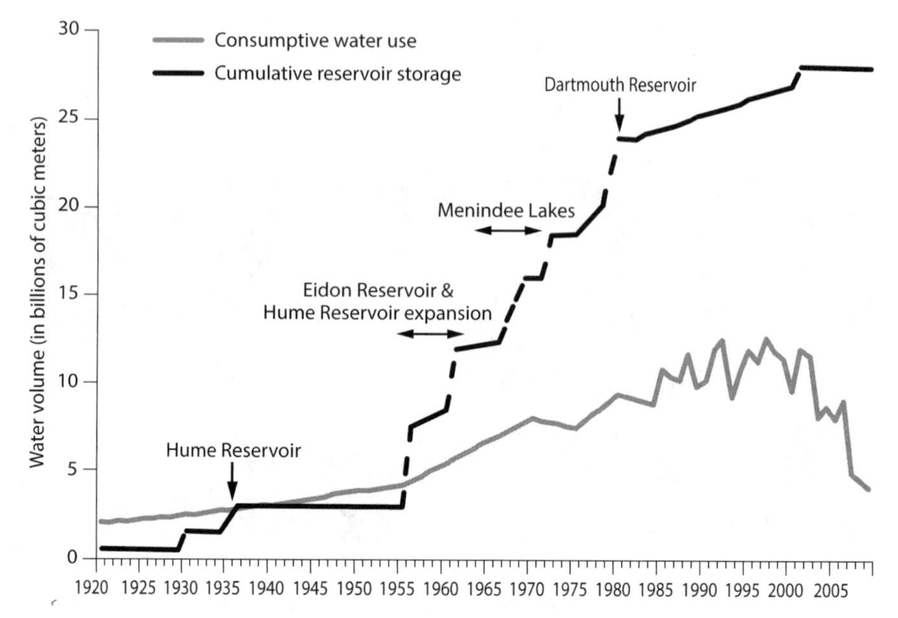

***Figure** 7-4. The construction of reservoirs in the Murray-Darling watershed accelerated rapidly from the 1950s until the end of the twentieth century, facilitating ever-greater use of water for irrigated agriculture. As a result, the flow of water in the lower Murray River has been reduced by 40 percent on average. During the Millennium Drought of 1997–2009, water consumption dropped precipitously because much less water was allocated for use during the drought.*

each state—were needed to avert conflict and facilitate agricultural development.

The River Murray Waters Agreement signed in 1915 helped to clarify how water would be shared among the three states, and also greatly facilitated interstate collaboration in building a network of dams, locks, and weirs to distribute water. By this time, each of the states had also begun instituting a system for allocating water among the individual water users within its boundaries. In this regard it is interesting to compare the approach taken to allocating water entitlements in southeastern Australia with the approach taken in the western United States. The timing and nature of water development in these two countries are uncannily similar. The discovery of gold in both places at nearly the same time, the mid-1800s, triggered mass migrations of fortune-seeking miners, along with a pressing need to produce food for the new arrivals in the goldfields. As

Table 7-1 Annual Water Budget for the Murray-Darling Watershed in Australia, in BCM (MAF)*

	Undeveloped Condition, No Water Use	Current (2008) Water Use Levels	Current Water Use Use with Projected 2030 Climate
Water Supply			
Inflows to rivers	28.6 (23.2)	28.6 (23.2)	25.8 (20.9)
Transfers into watershed	0	1.1 (0.9)	1.0 (0.8)
Return flows, irrigation & urban	0	0.2 (0.1)	0.2 (0.1)
Subtotal of Supply Inputs	28.6 (23.2)	29.9 (24.2)	27.0 (21.8)
Consumptive Uses			
Agricultural (irrigation & stock watering)	0	9.7 (7.9)	9.4 (7.7)
Urban	0	0.3 (0.2)	0.3 (0.2)
Losses in constructed channels & pipes & to groundwater pumping	0	1.4 (1.1)	1.4 (1.1)
Evaporation from reservoirs or natural lakes	4.4 (3.6)	3.9 (3.2)	3.5 (2.8)
Natural river losses (evaporation & seepage)	13 (10.5)	9.9 (8.0)	8.9 (7.3)
Subtotal of Consumptive Uses	17.4 (14.1)	25.2 (20.4)	23.5 (19.1)
Outflow from Watershed	11.2 (9.1)	4.7 (3.8)	3.5 (2.8)
Outflow as Percentage of Inflow	39%	16%	13%

*BCM = billion cubic meters; MAF = million acre-feet.

Water budget estimates based on Commonwealth Scientific and Industrial Research Organization, *Water Availability in the Murray-Darling Basin* (Canberra: CSIRO, 2008).

competition and conflict over scarce water supplies arose in the mining regions of both Australia and the United States, however, the two countries took very different approaches in defining the rights to use water.

In the western United States, a prior-appropriation system was adopted, as discussed in chapter 5. That system doled out water entitlements according to when a miner or farmer first began using water, thereby creating a chronological lineup of water-use priorities. When water was in short supply, those at the back of the priority list would receive no water at all.

The Australians approached water allocation in a very different way. They created a system of water entitlements in which each class of entitlement—that is, high-security or low-security water rights—would receive an adjustable allocation of water, dependent on overall water availability. In some years all water entitlement holders receive the full volume of their entitlements, but in drier years they receive only a portion of their entitlements. As Daniel Connell has put it, the Australian approach "tends to share the impact of shortages between entitlement holders rather than privileging one as opposed to another, thereby reflecting the cultural importance assigned to equity of sacrifice."

Unfortunately, freshwater ecosystems have been left short of water under both the Australian and the American systems of water allocation.

Nature Strained to the Breaking Point

The natural ecosystems of the Murray-Darling watershed have always experienced dramatic changes between wet and dry years, a phenomenon that Australian scientists call a "boom-and-bust ecology." When rain is plentiful and the rivers rise out of their banks and onto their vast floodplains, it's "boom" time (fig. 7-5). Thousands of wetlands scattered across the watershed turn from brown to green, plants burst into flower, fish populations swell, and millions of waterbirds breed in the marshes. Then, during the "bust" years—the interludes between wet periods—the natural ecosystems slip into quiescence, and the adaptive physiologies of the native species, honed over thousands of years of evolution, are put to the test.

There has been a lot more busting than booming taking place during recent decades. The construction of huge reservoirs during the twentieth century—providing greatly increased capacity to capture river water for use in irrigation—caused river flows to dwindle and largely curtailed the natural floods that are so important to the ecological health of rivers and wetlands in the watershed. By the late 1980s, 40 percent less water was reaching the river mouth near Adelaide. Fish populations had declined

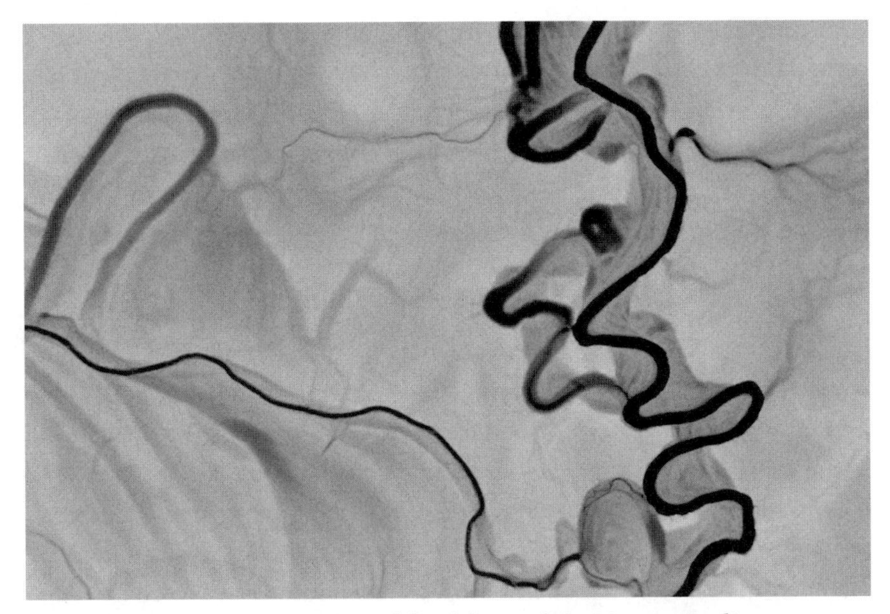

Figure 7-5. *The rivers, streams, and floodplains of the Murray-Darling watershed form an intricately patterned mosaic of water and land as seen from the sky. When water is plentiful, the rivers spill out of their channels and into a maze of smaller streams and ponds that appear as gray channels in this photo. The Murrumbidgee River, a tributary to the Murray River, is visible as a sinuous black thread on the right side of this image. During the highest floods, this entire landscape is underwater. (Image provided by Murray Scown.)*

by 90 percent since European settlement. The ecological health of the Murray-Darling system was about to have a breakdown.[5]

One of the early warnings of pending disaster was a buildup of salts in the floodplain soils, caused by the overwatering of crops. Excess irrigation water drained into the groundwater below, which then rose to the ground surface, bringing natural salts in the soil along for the ride. Lacking regular floods to flush the salt away, the floodplain soils became too saline to support native vegetation, such as the majestic river red gum trees that line the Murray River's banks. Similarly, agriculturally applied nutrients and other chemicals began accumulating in the Murray-Darling's waterways instead of being regularly flushed to sea. Those nutrients stimulated undesirable algal and plant growth and turned streams and wetlands into mucky slime.

Then, in 1991, the lower thousand kilometers of the Darling River turned into a toxic, stinking, slimy ooze of blue-green algae, killing just about everything that lived in the river, tainting public drinking water supplies, and causing the state of New South Wales to declare a state of emergency. It was time to take bold action to change the way water was being managed in the Murray-Darling watershed.

Capping Water Consumption

The toxic algal bloom in the Darling River became a catalyst for sweeping water reform in the Murray-Darling Basin in the mid- to late 1990s. However, as with most water policy reforms, this one can best be understood in the context of smaller changes that preceded it in earlier decades.

Mounting concerns over competition for water, along with widespread declines in ecological health, caused South Australia to stop issuing new water entitlements in the late 1960s and New South Wales to do the same in the early 1980s. By 1985, the states sharing the Murray-Darling and the federal government—joined together in a new Murray-Darling Basin Ministerial Council—were discussing the need to curtail issuance of new water entitlements throughout the watershed.

The states' efforts to shut down new water uses had not arrested the growing level of consumptive water use in the watershed, however. Many farmers still held entitlements to more water than they had been able to use previously. Throughout the 1980s and into the early 1990s, they continued to expand their farm production, using more and more of the water in their entitlements. Consumptive water use hit a new high in the early 1990s, triggering the Darling River algal bloom.

In 1993, the ministerial council called for a basin-wide audit of water use, published in 1995 after 2 years of intensive study.[6] The audit found that consumptive water use had grown by 8 percent since 1988. Most concerning was the realization that nearly one-third of the volume of authorized entitlements still remained unused; the audit projected that half of that unused water would likely be put to use in the near future, placing even more strain on the entire system. In response, the ministerial council called for a basin-wide limit on the allowable volume of consumptive use. "The Cap," as it came to be known in Australia, was implemented in 1997. It set a new limit on total maximum water consumption in the watershed at 11 BCM (8.9 MAF) per year, a level approximating the volume being consumed in 1993–94.

The Cap was a wake-up call heard throughout the Murray-Darling watershed. Farmers and towns would need to find a way to live within these newly imposed water constraints. In addressing complaints from farmers and towns aspiring to expand their use of water, the ministerial council was quick to point the way forward: "The Cap should restrain water diversions, but not development. With the Cap in place, new developments should be allowed, provided that the water for them is obtained by improving water use efficiency or by purchasing water from existing developments."[7] In other words, the users of the Murray-Darling water account had entered a new era that would feature water efficiency, conservation, and water trading.

More Crop per Drop

With the Cap in place and a need to accelerate improvements in water use to sustain continued growth in agricultural revenues, the state governments decided that it was time to privatize some of the formerly government-owned irrigation districts, based on a widely held belief that private companies could run the districts more efficiently and profitably. As used in this context, "improving efficiency" means using less water to produce the same agricultural output—more crop per drop.

The New South Wales government privatized five of its irrigation districts in 1995. It was a move that was well received by many farmers, as they viewed it as an opportunity to gain more control over their operations by becoming shareholders in an irrigation company. But they soon realized that their newly acquired real estate was in need of repair.

"The farmers quickly discovered that it was a very inefficient system they were operating," observed Perin Davey, the water policy officer for Murray Irrigation Ltd, the largest privately owned irrigation company in Australia. "They noticed there were a lot of inefficiencies when they had to pay for every megaliter that came down the channels, whether they used it or not. So the business had to look at how it operated and become more efficient. When we were privatized in 1995, we were operating at about 72 to 75 percent efficiency. Our latest assessment shows that we're now operating at over 85 percent, with the intent to keep improving further."

Those efficiency improvements are the outcome of an important deal worked out between the New South Wales government and the new irrigation companies during privatization negotiations. The state government committed to paying $190 million over 15 years to the irrigation companies to help increase irrigation efficiency and reduce salt problems, provided that the corporations would invest another $750 million over 30 years.[8] Davey's company, Murray Irrigation Ltd, used some of that money to upgrade its irrigation canals and make operational improvements, saving 30 million cubic meters (24,000 acre-feet) per year. Other irrigation companies and farmers' associations were quick to follow. The Ricegrowers Association of Australia, for example, administered a program that funded 65 different irrigation efficiency projects along the Murray River and one of its main tributaries, yielding 11 BCM (10 MAF) of water savings.

Government Funding for Water Buybacks

The imposition of a cap on the total volume of consumptive use in 1997 was a critically important step in arresting the depletion of river flows and associated ecological degradation. However, as the 1995 water audit had revealed, already authorized entitlements still exceeded the cap by as much as one-third, and consumptive use was repeatedly exceeding the cap limit of 11 BCM (8.9 MAF) (see fig. 7-6; also fig. 7-4).[9] The state and federal governments therefore faced a very difficult quandary: they needed to get that excess entitlement off the books, but they did not want to force reductions through purely regulatory means, or without appropriate compensation to affected entitlement holders.

The chosen solution was to buy the water back from willing sellers. The Murray-Darling Basin Ministerial Council launched an ambitious initiative in 2002 called The Living Murray. The ministerial council committed $700 million to the program over 5 years to purchase 500 million cubic meters (400,000 acre-feet) of water entitlements, to be held by the Commonwealth government and used for environmental purposes.

Purchases of water entitlements for environmental purposes have similarly taken place in the western United States for decades, but the way in which the purchased water is used in the United States differs from the

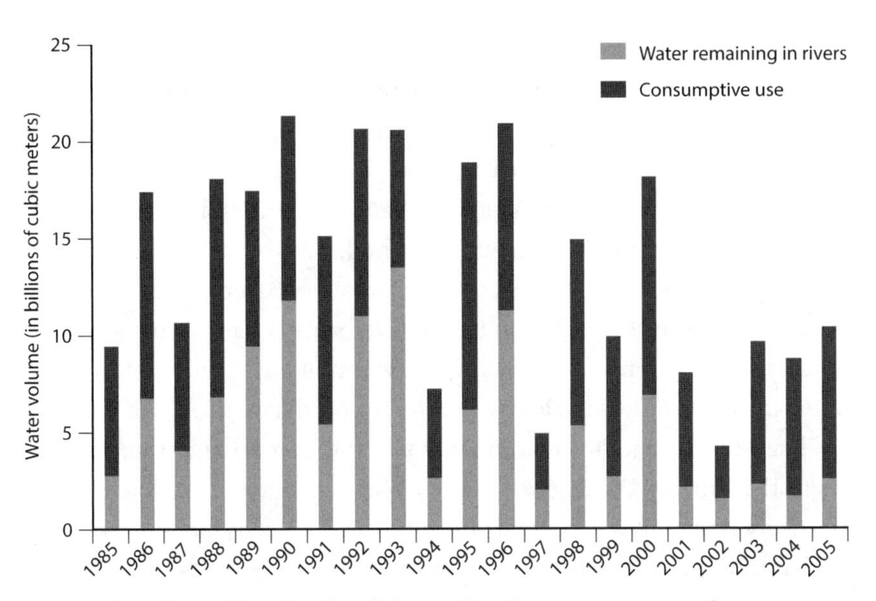

Figure 7-6. *The total height of each bar in this diagram represents the sum of all water that was available in rivers and streams in the Murray-Darling watershed of Australia over two recent decades. Much of the available water was consumptively used for farming, with the remainder flowing downstream.*

use of environmental water in the Murray-Darling. In the United States, when a water entitlement is purchased from a farmer or other water user for environmental purposes, the water is simply left "instream" by the original entitlement holder, and the water passively flows downstream as environmental flow. In contrast, in the Murray-Darling watershed, much of the environmental water is managed in the same way that a farmer would irrigate a crop: The water is typically stored in a reservoir prior to use, then carefully conveyed to a targeted wetland, floodplain forest, or stretch of river, oftentimes with the use of irrigation canals, pipes, and weirs to move the water. This enables a holder of an environmental water entitlement, such as the Australian government, to strategically direct the purchased environmental water to the locations where it is believed it can have the greatest ecological benefit. The Living Murray water purchases, for instance, were directed to six sites along the Murray River that are highly valued for their ecological attributes and their cultural and heritage significance to Aboriginal people.[10]

But Then It Stopped Raining

Taken in sum, the water reforms embarked upon in the late 1990s in the Murray-Darling watershed—establishment of a cap on consumptive uses, privatization of irrigation districts, and water buybacks—represent some of the most ambitious water policy changes attempted by any government anywhere. With those reforms, the basin's water managers took a giant leap forward toward more conservative, secure, and sustainable water use. But in the backdrop of these reforms, a dark cloud—or more accurately, an absence of rain clouds—was building, ready to test water users and governments in a way that none could have imagined.

The Millennium Drought snuck up on the farmers and water managers in the Murray-Darling watershed. The first signs of trouble came in 1997, when river flows dropped precipitously. The rains that year were not all that much less than normal, but even small changes in rainfall can have surprising impacts on a water budget—and particularly on river flows—in arid regions like southeastern Australia. During dry years, crops and natural vegetation will require more water than usual, owing to high temperatures and evaporation rates. With less rainfall available and higher water loss to crops and other plants, much less runoff makes its way through the watershed into the river. As a result, in the Murray-Darling, decreases in rainfall of less than 5 percent can reduce river flows by more than 20 percent.

During dry years, water will also evaporate more quickly from reservoirs, at the same time that farmers are pulling more water from the reservoirs for irrigating their crops. Compounding these problems is the fact that rivers flowing into the reservoirs are also low during dry times, their water flows insufficient to replenish the loss of water from the reservoir to evaporation and irrigation, causing reservoir levels to drop rapidly. All of these factors combine to create a very scary prospect: crops needing more water at a time when rivers and reservoirs are low.

The rains of 1998–2000 were closer to normal, and reservoir storage rebounded to a modest degree.[11] But then, in 2001, it seemed as though the rain had stopped altogether. With every month, the landscape became drier, and drier. Rainfall during 2004–2006 was only 16 percent lower than average, but averaged river flows across the watershed were much worse,

***Figure** 7-7. Thousands of mature river red gum trees lining the Murray River, many of them hundreds of years old, died during the Millennium Drought.*

at 39 percent less. In some of the southernmost areas of the watershed, the rivers had dropped by nearly 60 percent.

The reservoirs were quickly being sucked dry by farmers trying to keep their crops alive with irrigation. By 2007, the biggest reservoirs were nearly drained. Dartmouth Reservoir's storage had dropped to only 10 percent of its capacity, Eildon Reservoir was down to only 5 percent, and just 1 percent of the storage in Hume Reservoir remained. As John Conallin had put it, for the farmers and the ecosystems of the Murray-Darling watershed, it looked "like it was coming to the end of it all."

The state and federal governments quickly commissioned a number of assessments in the heart of the drought, to take stock of the damages that were accruing. The Sustainable Rivers Audit conducted in 2007 found only one of the twenty-three river ecosystems in the watershed to still be in good ecological shape, with twenty being rated "poor" or "very poor."[12] Along the Murray River itself, four-fifths of the beautiful river red gum trees—many of which were hundreds of years old—were showing stress, and thousands of those trees subsequently died (fig. 7-7), along with hundreds of Murray cod.

Things were very bad and getting worse on the farms. Due to a lack of rain and depletion of reservoirs, farmers were on average receiving only one-third of the water that they had been using before the drought, and some farms with lower-priority entitlements received no water allocation at all during the worst 3 years. Overall, dairy production in the watershed fell by an average of 14 percent, cotton fell by a fourth, meat by half, and rice farming stopped almost entirely.[13]

With the damage toll rising and with no end of the drought in sight, the federal government decided it was time to wrest control from the states.

The Heavy Hand of Government

"The drought kept going," remembers Perin Davey of Murray Irrigation Ltd. "Water got tighter and tighter. The storages were drying up. The water managers were struggling to meet critical human needs. They had to shuffle priorities to make sure they could get water to the towns."

Mary Harwood, of the federal government, said, "It was super scary. We were staring dreadful outcomes in the face, impossible choices about who gets the water."

The federal government was growing increasingly anxious and impatient with the states, pressing them to take more drastic action, and quickly. But then the negotiations fell apart. Davey recalls, "Towards the end of 2006, the state of Victoria walked away from the table and said, 'Sorry, we've got to look after our own state's interests.' And then all the others states followed. Malcolm Turnbull was the water minister at the time. He is a very good businessman, and a hardline negotiator. He decided that if he couldn't get a happy merger among the states, then the Commonwealth would go it alone. To do that, he brought in the Water Act."

The Australian Water Act of 2007 fundamentally shifted the balance of power among state and federal governments in water policy making. The federal government called on two international conventions, to which Australia is signatory, as the basis for taking charge. The Convention on Wetlands of International Importance, also called the Ramsar Convention, named for the city in Iran in which it was adopted in 1971, allows signatory countries to register specific wetland areas for international recognition, but then those countries must commit to their protection. Similarly, Australia is signatory to the Convention on Biological Diversity, adopted during the Earth Summit in Rio de Janeiro in 1992. These international agreements, the Australian government asserted, required the country to prevent further degradation of the freshwater ecosystems and species in the Murray-Darling watershed. And that meant making serious adjustments in the way that water was being used.

The Water Act created a new Murray-Darling Basin Authority and directed it to prepare a Basin Plan for the integrated management of water resources throughout the watershed. In concert with the Water Act, the Australian government announced a new national initiative, Water for the Future, and committed $12.6 billion over 10 years to its implementation. The funds were earmarked to deliver $5.8 billion for rural water use and infrastructure projects to improve the efficiency of water use on farms, and $3.1 billion more for water buybacks.

Drafting the Basin Plan

The volume and detail of analysis that were undertaken to prepare the Basin Plan during the 3 years following creation of the Water Act is almost beyond comprehension. The technical experts building the plan divided the basin into nineteen different watersheds for assessing surface water and twenty-three more for examining groundwater in aquifers. Designated within those forty-two water planning areas were 107 subareas in which water-use limitations, or "sustainable diversion limits," would be set. Environmental scientists analyzed information about more than 2,400 different "environmental assets," ranging from small alpine streams to large lowland rivers. Teams of economists and sociologists were enlisted to evaluate the potential socioeconomic consequences of alternative scenarios for limiting water use. The resulting report, the *Guide to the Proposed Basin Plan*, along with its many appendices and supplementary reports, represents a giant repository of information on the Murray-Darling watershed.[14] Over the course of the 3 years of its preparation, hundreds of staff were deployed in its formulation, at the cost of many millions of dollars.

When the *Guide to the Proposed Basin Plan* was publicly released in October 2010, the Australian government was unprepared for the adverse reaction that erupted from rural towns and irrigated farms. But they could have anticipated some hostility. The guide called for a minimum reduction of 3 BCM (2.4 MAF) in the existing level of consumptive use basin-wide. Much of the rollback in water use was predicated on the need to protect at least 60 percent, and preferably 80 percent, of the natural flow in all rivers and streams in the watershed to restore their ecological health. Because of differing levels of existing use in subwatersheds, and given recent efforts to improve irrigation efficiencies and buy back water for the environment, the size of the targeted cutbacks would affect different parts of the watershed to differing degrees. The reductions suggested in the guide averaged 30 percent overall, but some areas of the watershed would be faced with water-use reductions of as much as 40 percent.

Upon reading that bad news, thousands of farmers threw the plan into bonfires as they protested in rural towns. Even though the water-use reductions would largely be accomplished through buybacks, many small rural communities had already been seeing the undesirable consequences of governmental water buybacks and wanted no more. Farms were going out of production, farmers were moving out of the area, and rural businesses were closing as they lost customers. Photographs of the bonfires, fueled with hundreds of copies of the *Guide to the Proposed Basin Plan*, quickly spread through the news media and Internet. Overnight, the Australian government became painfully aware that the people most affected by the plan—farmers and townspeople in irrigated areas of the watershed—had not been adequately consulted or engaged in the plan's formulation.

Upon reflection, Mary Harwood said, "We learned some hideous lessons from the release of the draft plan. It was an impossible task, you know, to get some sort of perfect rendition of a sustainable allocation across all of those [watersheds], and all the tradeoffs involved. I think when the [Murray-Darling Basin] Authority was faced with that gargantuan task, they decided they needed to go into a bunker and beaver away from a science basis as best they could, and then sort of emerge with a draft."

Perin Davey explained the rural reaction to the guide. "The community was pretty much in the dark [during the plan's development]. The community was told, 'We're doing these basin plans, don't you worry about it.' But the guide came out and it was an absolute shock to those communities that had been told, 'Don't worry about a thing; this will be good for you.' We had no idea how much water they [the federal government] were going to try to recover, or how they planned to do it. The guide is a classic case study of what not to do. Putting it out on the table and then 2 days later rocking up in some town saying, 'Here it is, aren't we good. Here's your 500 pages, here's your answer.' Those farm communities had just gone through 10 years of turmoil and drought, financially hung by the teeth. They were just so unprepared for what was handed to them."

The Dust Slowly Settles

John Conallin had returned from graduate school in Denmark in 2009, just in time to witness the social maelstrom set off by the release of the *Guide to the Proposed Basin Plan*. He met Dave Leslie, the general manager of the Murray Catchment Management Authority (CMA), soon thereafter. The CMAs were set up in New South Wales in the 1990s to provide assistance to farmers and other rural residents interested in learning more about natural resource management. The Murray CMA had not been deeply involved in water issues previously, but with water controversy swirling in the region, Leslie was anxious to advance a new approach, "a model that includes people in the landscape and stops treating environmental problems just as biophysical problems to be solved by managers in Sydney and Canberra. After all, people caused the problems, so people have to be part of the solution." When Leslie met Conallin, he knew that he had found someone who could help advance that idea.

Conallin remembers that he and Leslie "got to talking and Dave asked me what I thought about using science to inform management, but including the social aspirations as well. He asked me if I could step out of a pure science role and into a facilitation and communication role with the community. That was the role I'd been waiting for. I just said, absolutely I can."

Conallin spent the first 6 months in his new job at the Murray CMA studying the science behind the draft water plan, and listening to the community. "There was almost hysteria in some ways. What I found really interesting was that they weren't just complaining about water cutbacks; it was the landholders that were yelling the loudest about the environmental state, they were the ones advocating for environmental flows. All the fish were dying, and these were fish that you'd fish a lifetime to catch, and to see twenty or thirty of them floating dead under the Deniliquin

***Figure* 7-8.** *I'm holding a Murray cod in this photograph, posing proudly with John Conallin after catching "the fish of a lifetime."*

Bridge. . . . I mean, there were fishermen that were so devastated, they saw their whole life disappearing with their kids'. They thought, 'My kid will never catch a Murray cod. They're all dead.' And you know, it was like they just gave up.

"I knew I just had to let them vent. Even though I was a local bloke, there was a lot of yelling. You've got to be able to handle that, and not everyone can. Some will react and say 'it's not *my* fault.' But that's not the point. They just need to blow it off. You know, we really are social animals. You have to break the ground, smell each other, suss each other out a bit, gain some trust, show you have stake in what is happening, and only then can you start to talk about things, as opposed to just coming in and saying, 'This is what we're going to do.'"

One of the first things Conallin did was to team up with a fish ecologist, Lee Baumgartner, and his colleagues at the Narrandera Fisheries Centre. They brainstormed about ways to get the local community members more involved in the decision making, and get them excited about science. They inserted electronic tracking tags into some of the Murray cod. When a pulse of water was released from an upstream dam for environmental purposes, the cod raced upstream to feed and spawn, and the scientists tracked them on their computers. When Conallin showed a video of the fish movements in the local community, he knew he had turned a corner in gaining their interest, and their trust.

Perin Davey emphasizes the importance of communicating environmental goals with local communities. "None of us around here wanted to break the environment. But you have to show us what you're doing with the water. You have to monitor the results and show us. As a taxpayer, I mean, this is $12 billion worth of taxpayer's money that's to be invested over 10 years' time, and we will want to know what it's achieved."

Looking Back, and Taking Stock of What Was Learned

Beginning in 2010, the skies above the Murray-Darling watershed opened up and started raining again, and it kept raining hard for 2 years. The rain, and time, has helped to ease the tensions. The Final Basin Plan was released in November of 2012, without much fanfare. I think the Australians were relieved in some way to have both the drought and the Basin Plan behind them.

The final plan relaxed the mandatory reductions in consumptive use called for in the draft plan by nearly 10 percent, from 3 BCM (2.4 MAF) to 2.75 BCM (2.2 MAF). This final target is obviously the result of a long and very difficult negotiation among the federal and state governments, as well as a response to the concerns expressed in the farming communities. The 2.75 BCM reduction in water-use entitlements will bring their total volume to just below 11 BCM (8.9 MAF)—about the same level that was called for in the 1997 Cap. Criticism will persist among many in the scientific community, because even the draft plan suggested that as much as 7.6 BCM (6.2 MAF) of reductions would be necessary to provide a high degree of confidence in ecological restoration. Only time, extensive and careful monitoring, and another big drought will tell whether or not the 2.75 BCM rollback is enough to sustain freshwater ecosystems and species.[15]

It is interesting to note that recent assessments of the economic impact of the Millennium Drought have documented that the farm economy overall weathered the drought fairly well. Agricultural water use during the drought dropped by two-thirds, because of the great reductions in allocations given to each water entitlement, but farm revenues dropped by only 20 percent.[16] This was largely made possible by huge increases in water productivity, that is, the amount of revenue generated per unit of water consumed. The amount of water used to produce each dollar of farm revenue dropped by a whopping 241 percent during the drought.

This increase in water productivity resulted from the widespread investment in irrigation efficiency and, more importantly, from water trading. Prior to 2006, only 10 to 20 million cubic meters (8,000 to 16,000 acre-feet) of water were being traded each year among different farming regions in the watershed. By 2008–2009, trading activity had increased to 500 million cubic meters (400,000 acre-feet) per year even though water prices had risen considerably. This buying and selling of water entitlements was extremely important to the farm economy because it allowed water to move from lower-valued to higher-valued crops. During the peak of the drought, huge volumes of water were sold by farmers producing rice, cereal grains, and cotton—where the economic returns on water use range from $0.3 million per BCM to $0.6 million per BCM—to farmers who could generate $1.8 million to $15 million with every BCM by growing grapes or other fruits, vegetables, or plants for sale at nurseries.

Water trading has given the Murray-Darling farmers options that can help them stay on the farm, and stay alive. When I talked with farmer Howard Jones (fig. 7-9) of Mildura—who loves wetlands and wild creatures as much as he loves growing grapes and drinking fine wines—he explained the importance of water trading to farmers: "As an irrigator, you get an annual allocation of water for your entitlements. Should we be in a dry period, and there's insufficient water in the dams to supply the full allocation, you'll get a partial allocation. That might necessitate buying some water on the market to supply your trees or vines, to top up beyond your own water entitlement. Fortunately, there will almost always be some farmers that will want to sell or lease their water rather than trying to make a crop that year. It works out for everybody that way."

Jamie Pittock, a water policy expert at the Australian National University in Canberra, points to another benefit realized during the drought: "We often say that you can't manage what you don't measure. One of the great side benefits of the water market reforms was much-improved water accounting. People aren't going to trade water unless they really know what they're getting or selling. Each of the four states and the capital

Figure 7-9. *Howard Jones, a wine grape farmer near Mildura, loves wandering around the Murray River's wetlands as much as he loves growing grapes. He was thrilled to show me one of his favorite wild creatures, a shingleback lizard (*Tiliqua rugosa*).*

territory had been counting water in somewhat differing ways, and that had to be harmonized and nationalized. The new accounting procedures will give us the information we need to do a better job of managing water in future."

Pittock also emphasizes that the best way to improve a water plan is to require that it be reviewed at regular intervals: "One of the best aspects of the Murray-Darling Basin Plan is that it has to be revised at least every 10 years, creating a regular opportunity for us to reflect on how we can do better in the period to come."

Based on the reforms and adaptive learning that has taken place in the Murray-Darling watershed, I would give the water users and policy makers in Australia very high sustainability scores (table 7-2). Virtually everyone I talked with in the country—from high-level water officials in Canberra to irrigators and fishermen in rural communities—agreed that more effort needs to be put forth to empower those in rural areas to have more influence over their future, helping them contribute to a shared vision for water management (principle #1).

Table 7-2 Sustainability Scorecard for Murray-Darling Watershed, Australia

Sustainability Principles	Little to No Progress	Notable Progress	Strong Performance
Principle #1: Build a shared vision for your community's water future.		√	
Principle #2: Set limits on total consumptive use of water.			√
Principle #3: Allocate a specific volume to each user, then monitor and enforce.			√
Principle #4: Invest in water conservation to its maximum potential.			√
Principle #5: Enable trading of water entitlements.			√
Principle #6: If too much water is being consumptively used, subsidize reductions in consumption.			√
Principle #7: Learn from mistakes or better ideas, and adjust as you go.			√

In the federal government halls of Canberra, the new buzzword is *localism*. As Dave Leslie emphasized during our conversation, "the senior bureaucrats and politicians need to learn to accept the risks of delegating downwards. They need to be comfortable in devolving resources and accountability to lower levels. And at those lower levels, the capacity has to be built, and that requires bringing in people like John [Conallin] with the technical and communications skills to get all the people together on the ground."

I was also quite heartened to hear people in rural communities acknowledge that those in the government offices in Canberra are trying to do better, trying to avoid the social mistakes made during the Basin Plan's formulation. "The consultation has to be really, really good," said Perin Davey. "To their credit, the Environmental Water Holder [the federal office responsible for environmental water buybacks] has learned an awful lot since they've started operating. They have started to build the relationships out in the country. And that's what a lot of this is about: it's building relationships."

CHAPTER EIGHT

Chasing Hope

I am regularly asked how I can remain optimistic in spite of what I know about the world's water challenges. There certainly is plenty of depressing news about water shortages being reported in the global news media. The Internet search engine that I have set up on my computer automatically delivers a continual stream of stories about water struggles around the world, each day bringing more bad news of people, economies, and ecosystems that are suffering under water scarcity.

In truth, an objective appraisal of the global water situation does not reveal a hopeful outlook. The trend is not good. Water shortages are beginning to occur in places that once seemed to have plenty of water, such as in the eastern United States and Southeast Asia, and the impacts of shortages appear to be intensifying everywhere. The volume of water use has pushed up against the limits of its availability in too many places.

But it does not have to be this way. I remain optimistic about our global water future for one primary, and rather ironic, reason: we presently manage water so poorly, and so wastefully, that plenty of opportunity exists to meet our needs, using available water supplies, for another two to three decades. The recent history of water use in the United States offers a case in point. Water withdrawals in the United States peaked in 1980, and have remained constant ever since, even while the U.S. population grew by more than a third. This has been accomplished primarily through improved water-use efficiencies in electricity production and irrigated agriculture, the two largest categories of water use in the United States. In some parts of the country, improved water efficiencies during the past three decades have enabled cities to grow without further depleting local water sources. In other places, water savings have reduced the pressure on strained water sources, lessening the risk of running out of water and restoring life-supporting water flows to withering freshwater ecosystems.

There is great potential to do a lot more with water conservation, both in the United States and in all other countries experiencing water shortages. As discussed in chapter 3, a reduction of just 15 to 20 percent in consumptive water use in agriculture would free up enough water to substantially alleviate water shortages in most water-stressed areas of the world. And every water-stressed city could further reduce its demands on depleted water sources by applying readily available, cost-effective urban water conservation measures. As the World Water Council emphasized in its *World Water Vision* report in 2013, "the crisis is not about having too little water to satisfy our needs. It is a crisis of managing water so badly that billions of people—and the environment—suffer badly."[1]

Over a longer time horizon, such as for the next 50 years, our ability to secure sufficient water for drinking and other domestic purposes, growing food, and generating electricity will depend largely upon the size of the global human population and on advancements in technology. More specifically, our water future will hang on the question of whether technological progress can get out ahead of population growth, such that the current trend of worsening water scarcity begins to reverse. Of greatest importance here is the inextricable linkage between water and energy use. Of the six "water tools" I describe in chapter 3, our utilization of three of those tools—water importation, water reuse (recycling), and desalination—is heavily constrained by their energy costs. Those costs will not be reduced much by further improvements in water recycling or desalting technology; instead, the prospects for broader use of these water tools will hinge on reductions in the cost of producing the energy required to operate these technologies, and doing that in a way that does not increase carbon emissions that are driving climate change.

If we can produce energy in a less costly and less climate-altering way, the long-range potential for technologies such as desalination—turning salty water into freshwater—could be quite remarkable. Consider the fact that 97 percent of all water on Earth is found in the oceans, and more than 40 percent of the global population already lives within 100 kilometers of a coastline. A massive shift toward the use of desalination by coastal cities could greatly lessen the strain on many of our planet's watersheds and aquifers. At the same time, however, we will need to be careful not to create ecological problems in our coastal habitats by improperly disposing of salt-concentrated brine, the by-product of desalination.

While I am optimistic about the immense potential for improving our use and management of water, and thereby our ability to alleviate water shortages, I am concerned that these improvements are coming far too slowly. As a consequence, more and more people are being affected by water shortages each year, and more freshwater ecosystems are being damaged from overuse unnecessarily. We must move swiftly to implement corrective measures.

We should not expect, or wait for, our governments to lead the way in resolving water scarcity, however. Many factors—as detailed in chapter 4—limit the ability, willingness, and agility of most governments to advance change rapidly. Instead, the leadership and inspiration for change in water management will almost always need to start with the individuals, organizations, and companies that are being adversely affected by water shortages. They have much at stake, and they possess the need, passion, and commitment to push for needed changes. This does not mean that they should act in exclusion of government, but they do need to lead, with the hope that their governments will follow.

Fortunately, there are millions of people—from water professionals to scientists to local community leaders—proposing practical water solutions and galvanizing support for their implementation. Their inspired efforts have brought depleted water sources, degraded ecosystems, and the communities that rely upon them back to life. There are many commonalities among these success stories, many of which I have highlighted throughout this book.

In just the past decade, I have seen some dramatic turnarounds in places where solutions seemed unattainable previously. In almost all of those cases, the shift in momentum was catalyzed by an individual or an organization that saw the potential for resolution, and inspired a movement in that direction.

A very hopeful story is now finally emerging in the Colorado River Delta, where a coalition of conservation organizations called the Colorado River Delta Water Trust is making strong progress under the leadership of

Yamillet Carrillo and Osvel Hinojosa. These two Mexican conservationists have known each other since they were teenagers in Monterrey, Mexico. During college, they were both recruited to help with scientific inventories of the Colorado River Delta's remnant wetlands. As they learned about the vast wetlands and wildlife that had been lost with the drying of the river, they each became motivated to find some way to bring water back into the delta. Both pursued graduate degrees at the University of Arizona in Tucson, where Carrillo received her doctoral degree in renewable natural resource management and Hinojosa earned his doctorate in wildlife and fisheries science. They are now a unified force for nature in the delta, working for the nongovernmental organization called Pronatura Noroeste, which has joined with the Sonoran Institute, the Environmental Defense Fund, and The Nature Conservancy in forming the Delta Water Trust. They are buying water rights from farmers and other water users to restore river flow through the desiccated delta. Their efforts recently got a boost when the U.S. and Mexican governments jointly committed to improving irrigation efficiency on Mexican farms and restoring water flows in the delta, in the hope of fostering a more sustainable future for the Colorado River. With the return of the water, many will be watching for the reappearance of the behemoth totoaba fish, and anxiously anticipating the return of the delta's legendary richness of waterfowl.

Water is already flowing again, at least occasionally, in the lower Tarim River in far northwestern China. During the latter half of the twentieth century, use of the Tarim's water for agriculture had increased rapidly, to the point of completely drying the lower river. Xiaoya Deng, a PhD student in the School of the Environment at Beijing Normal University, grew up in the Tarim watershed. She remembers visiting the river—known as the "mother river" of Xinjiang Province—with her father on numerous occasions as a child, during a time when plentiful water flowed downstream. But when she visited the river again as an adult in 2009, she was dismayed to see that all of the majestic cottonwood trees (*Populus euphratica*) along the lower river had died or were heavily stressed from lack of water. Speaking of her beloved river, Deng said, "There was no green there. I really wanted to do something good for her."

Deng and her university colleagues have amassed compelling evidence that the cottonwood forest could be restored if water flows could be returned, even intermittently, to the river. In response to this scientific information, the Chinese government invested $1.75 billion in an "ecological water conveyance project" designed to restore the ecological health of the Tarim River and its floodplain forests. Deng's research has helped government officials understand how the occasional, timely release of water from upstream reservoirs could bring the forest of cottonwood trees back to life. At least thirteen pulses of water have been released into the lower Tarim under the restoration program, and the cottonwoods have responded with spectacular new growth. Government funding has enabled many of the farmers using the Tarim's waters to install water-efficient drip irrigation equipment, thereby reducing the volume of water needed on farms so that more can be left in the river. Deng is quite proud to share her reborn river with visitors, particularly in October, when the cottonwood trees are glowing in their fall colors.

Some nongovernmental organizations and private companies are advancing water strategies that are bringing relief to many places simultaneously. Andrew Warner, one of my colleagues at The Nature Conservancy, was born in Tripoli, Libya, and thanks to his father's enlistment in the U.S. Air Force, lived in seven different parts of the world before he graduated from high school. Warner's familiarity with military organizations was surely a factor in his proposal that The Nature Conservancy form a partnership with the U.S. Army Corps of Engineers in their management of dams across the United States. The Corps of Engineers is one of the world's largest water management agencies, with responsibility for operating nearly 700 dams and much other water infrastructure in the United States. In 2002, The Nature Conservancy and the Corps entered into a nationwide partnership agreement called the Sustainable Rivers Project, through which they are exploring ways to improve dam operations to benefit ecological health, recreational opportunities, flood management, water supplies, and production of electricity through hydropower generation.

Private companies, too, can be catalysts for change. The Coca-Cola Company, for example, in 2007 launched a global Replenish program that is making more water available to local communities and freshwater ecosystems by investing in local community water projects, such as drilling new groundwater wells in poor rural communities or reforesting denuded watersheds to improve their water flows. By the end of 2012, the company had provided $247 million in support of 386 different projects in ninety-four countries.[2]

Other companies are similarly engaging with local communities on water issues as an integral feature of their business models. In 2013, the Stockholm Industry Water Award—awarded each year as part of World Water Week in Sweden—was given to Netafim, a leading worldwide provider of equipment for improving irrigation efficiency. The company has specialized in developing water solutions appropriate for poorer farmers in developing regions of the world. Beyond their manufacture of irrigation equipment, Netafim provides training to farmers in developing countries, helping them maximize their crop yields while minimizing their water consumption. In Niger, for example, application of the Netafim irrigation system enabled farmers to more than triple their production of fruits and vegetables, and in India, more than 40,000 small-scale farmers increased their incomes by an average of 20 percent over 3 years.[3] Corporate programs such as this one, which provide training along with improved technology, will be critically important to alleviating water shortages and improving food production in poorer regions of the world.

One of the more hopeful indications I have witnessed in the past couple of years is a growing understanding within the private sector of the need to help strengthen government's abilities to manage water well, thereby providing greater certainty of water supplies. The United Nations, through its CEO Water Mandate program launched in 2007, has been facilitating a productive dialogue among hundreds of national and multinational corporations, investment and development banks, nongovernmental organizations, academic institutions, and other interested parties.[4] The Water Mandate members "seek to make a positive impact . . . by mobilizing a critical mass of business leaders to advance water sustainability solutions, in partnership with the United Nations, civil society organizations, governments, and other stakeholders." The participants in Water Mandate discussions have explored the boundaries of corporate responsibility in helping to resolve local and regional water crises, and have worked together to develop guidelines for corporate participation in water governance.

Many of the companies that are starting to engage with other water users and governments to improve water governance are quickly finding out that no single company or organization—no matter how large or resource rich—will likely be able to fix recurring water shortages acting alone. Instead, companies and civil society organizations have come to realize that only by acting together in collective, collaborative fashion, as the stakeholders at Lake Naivasha have, can they marshal the capacities, resources, and expertise to overcome local water problems.[5] An online tool called the Water Action Hub has been developed by the Water Mandate for companies wishing to identify potential collaborators sharing the same water source. If these companies can find ways of collaborating productively, both among their peers in private business and with governmental authorities and civil society organizations, they could be of great benefit in places where the government has been unable or unwilling to implement necessary corrective actions to resolve water shortages. To foster long-lasting water success, it will be essential that these corporate actors focus on three particular activities: (1) they must learn the language of water budgets and strategically contribute resources toward cost-effective and practical solutions that can rebalance overdrawn budgets; (2) they must help empower local communities of water users to shape their collective water futures; and (3) their work should supplement, *and never supplant*, the role of governments in managing water. Their ultimate goal should always be to build the capacity of governments and local communities to manage water well.

These stories are but a few of the hundreds of hopeful signs emerging around the world today. Too many communities have been chasing water, for too long. It is time to start living within the limits of water's natural availability so that we can reap the benefits of a water-secure future. Because only when we each have enough will we be able to truly appreciate the miraculous fortune of being born on the one blue planet.

Later that night it began to rain. Then to pour—as if the skies were saying that as soon as we all managed to cooperate among ourselves, then one way or the other there would be enough water for everyone.

Stanley Crawford, Mayordomo: Chronicle of an Acequia in Northern New Mexico

Acknowledgments

Many friends and colleagues helped with the writing and production of this book, and I am most grateful to them all. I am especially thankful to Bill Ginn and The Nature Conservancy for granting me some time away from my regular work duties to work on this book, and to my wife, Martha, and son, Henry, for their patience and understanding while I was consumed with its writing.

The entire Island Press team has been wonderfully supportive. They immediately recognized the book's potential to help educate people around the world about water scarcity issues and solutions. I am particularly indebted to Barbara Dean of Island Press and to Martha Hodgkins for their insightful editing suggestions.

Marcia Rackstraw created the illustrations for the book, and Emily Powell produced the maps. Great thanks for their wonderfully creative ideas and approaches.

I am forever indebted to David Harrison for his mentorship and for helping me to see water budgets as the essential foundation of all water problem solving.

The book has been reviewed in whole or in part by many people. Stuart Orr and Tony Maas offered wise advice on water governance issues. Brad Udall helped me to understand the details of the Colorado River Compact. Emily Powell offered excellent suggestions and legal research for strengthening the whole book, as did K. J. Joy, John Kinch, Madeline Kiser, Cindy Loeffler, Jamie Pittock, Jessica Gephart, and Melissa Duvall.

Many Australians opened their homes and minds for me while I was gathering information for my chapter about the Murray-Darling Basin. Jamie Pittock was a generous and capable host, tour guide, and policy interpreter. John Conallin shared rich perspectives about farmers and rural communities and arranged for a memorable fishing trip on the Edward River with Ian Fisher. I learned a great deal from interviews with Perin

Davey, Mary Harwood, David Papps, Howard Jones, Deb Nias, David Leslie, Digby Jacobs, Barry Hart, Peter Draper, Louise and Andrew Burge, and Elizabeth Stott. Joy and Stewart Scott graciously hosted me on my travels, as did Howard and Jill Jones, John and Jemmah Conallin, and Jamie Pittock and Cath Webb. Francis Chiew, Andy Close, Geoff Podger, and Tom Rooney shared useful data about the river.

Both Richard Fox and Stuart Orr provided interviews and information for my story about Lake Naivasha in Kenya. My story about the Guadalupe–San Antonio River system and the regional planning process in Texas was richly informed by interviews with Cindy Loeffler, Robert Mace, Carolyn Britton, Con Mims, Ryan Smith, Laura Huffman, Chloe Lieberknecht, and Kirk Winemiller.

Last but certainly not least, a big thank you to Lindsay Boring and Denise McWhorter of the Jones Ecological Research Center for providing a comfortable writing refuge in the wilds of south Georgia, to Dr. Mohit Nanda for repairing my detached retina so that I could see my way to the end of this project, and to the baristas at the Mudhouse in Crozet for keeping me well caffeinated and productive.

In addition to help from those mentioned above, I was fortunate to have several especially helpful sources of graphics and other information:

Figure 1-1 (map of Colorado River watershed): Map layers were provided by National Oceanographic and Atmospheric Administration (NOAA) (Basins of the Colorado Basin shapefile, U.S. Cities shapefile, Rivers of Colorado Basin shapefile); Esri (World Countries shapefile, U.S. States shapefile); DIVA GIS (USA Inland Waters shapefile); National Atlas (Streams and Waterbodies shapefile, Dams shapefile); and University of Arizona Institutional Repository (UAiR) (Central Arizona Project shapefile).

Figure 2-5 (water budget of the Colorado River watershed): Data used in this diagram are from Joan F. Kenny, Nancy L. Barber, Susan S. Hutson, Kristin S. Linsey, John K. Lovelace, and Molly A. Maupin, *Estimated Use of Water in the United States in 2005* (Reston, VA: U.S. Geological Survey, 2009) and U.S. Bureau of Reclamation, *Colorado River Basin Water Supply and Demand Study*, associated technical reports (Denver, CO: USBR, 2012).

Figure 2-6 (water level changes in High Plains Aquifer): Data used in this map are from Virginia L. McGuire, *Water-Level and Storage Changes in the High Plains Aquifer, Predevelopment to 2011 and 2009–11* (Washington, DC: U.S. Geological Survey, 2013). Map layer of U.S. states is from Esri.

Figure 3-2 (hydrographs from Santiago River, Mexico): Data are from Arjen Y. Hoekstra and Mesfin M. Mekonnen, *Global Water Scarcity: The Monthly Blue Water Footprint Compared to Blue Water Availability for the World's Major River Basins* (Delft: UNESCO-IHE Institute for Water Education, 2011).

Figure 7-1 (map of Murray-Darling Basin, Australia): Data layers were provided by Global Runoff Data Centre of Germany and the Commonwealth of Australia, Department of Sustainability, Environment, Water, Population and Communities.

Table 1-1 (world's most depleted freshwater sources): This list is based on output from the WaterGAP model developed at the University of Kassel in Germany; Arjen Y. Hoekstra and Mesfin M. Mekonnen, *Global Water Scarcity: The Monthly Blue Water Footprint Compared to Blue Water Availability for the World's Major River Basins* (Delft: UNESCO-IHE Institute for Water Education, 2011); and Tom Gleeson, Yoshishide Wada, Marc F. P. Bierkens, and Ludovicus P. H. van Beek, "Water Balance of Global Aquifers Revealed by Groundwater Footprint," *Nature* 488 (2012): 197–200.

Table 2-1 (water withdrawals and consumptive use): Withdrawal estimates included in this table were based on Joan F. Kenny, Nancy L. Barber, Susan S. Hutson, Kristin S. Linsey, John K. Lovelace, and Molly A. Maupin, *Estimated Use of Water in the United States in 2005* (Washington, DC: U.S. Geological Survey, 2009) but updated with estimates for thermoelectric and irrigation withdrawals taken from Glenn D. Schaible and Marcel P. Aillery, *Water Conservation in Irrigated Agriculture* (Washington, DC: U.S. Department of Agriculture, 2012); Igor A. Shiklomanov, "Appraisal and Assessment of World Water Resources," *Water International* 25 (2000), 11–32; Charles J. Vorosmarty and Dork Sahagian, "Anthropogenic Disturbance of the Terrestrial Water Cycle," *BioScience* 50 (2000), 753–765; Mesfin M. Mekonnen and Arjen Y. Hoekstra, *National Water Footprint Accounts: The Green, Blue and Grey Water Footprint of Production and*

Consumption (Delft: UNESCO-IHE Institute for Water Education, 2011). Consumptive use estimates for the U.S. are from Electric Power Research Institute (EPRI), *Evaluating Thermoelectric, Agricultural, and Municipal Water Consumption in a National Water Resources Framework* (Palo Alto: EPRI, 2013).

Notes

Chapter 1

1. Marc Reisner, *Cadillac Desert: The American West and Its Disappearing Water* (New York: Penguin, 1993).
2. Phillip Fradkin, *A River No More* (Berkeley: University of California Press, 1996).
3. An acre-foot of water is the volume of water required to inundate 1 acre of land to a depth of 1 foot.
4. Most accounts of the Colorado River Compact report that only 15 million acre-feet were allocated by the compact, with the upper and lower basin states sharing this volume equally (each would receive 7.5 million acre-feet). However, the compact actually allocated another 1 million acre-feet to the lower basin states, assuming that this additional water would come from the Gila River basin and not from the flows passing downstream from Lee's Ferry.
5. A recent summary of climate studies for the Colorado River basin is available in the *Colorado River Basin Water Supply and Demand Study* (Denver: U.S. Bureau of Reclamation, 2012).
6. Sally Adee and Samuel K. Moore, May 28, 2010, "In the American Southwest, the Energy Problem Is Water," IEEE Spectrum, http://spectrum.ieee.org/energy/environment/in-the-american-southwest-the-energy-problem-is-water/1.
7. Barbara Morehouse, George Frisvold, and Rosalind Bark-Hodgins, "How Can Tourism Research Benefit from Multi-disciplinary Assessments of Climate Change? Lessons from the U.S. Southwest," in *Developments in Tourism Climatology*, eds. Andreas Matzarakis, Christopher der Freitas, and Daniel Scott (Freiburg: International Society of Biometeorology, 2007), 274–281.
8. "The Colorado River Delta Blues," *Los Angeles Times*, March 25, 2012, accessed December 5, 2013, http://articles.latimes.com/2012/mar/25/opinion/la-oe-postel-colorado-river-delta-20120325.
9. Mark K. Briggs and Steve Cornelius, "Opportunities for Ecological Improvement along the Lower Colorado River and Delta," *Wetlands* 18 (1998): 513–529.
10. "Murray-Darling Basin Plan," video posted by Minister for Sustainability, Environment, Water, Population and Communities Tony Burke, December 6, 2012, www.youtube.com/watch?v=Wumfo3AJ57c.
11. See for example Sandra Postel and Brian Richter, *Rivers for Life: Managing Water for People and Nature* (Washington, DC: Island Press, 2003).
12. Joanne Yao, *Iraq's First National Park: A Story of Destruction and Restoration in the Mesopotamian Marshlands*, Circle of Blue, September 4, 2013, www.circleofblue.org/waternews/2013/world/iraqs-first-national-park-a-story-of-destruction-and-restoration-in-the-mesopotamian-marshlands/.
13. Jay Famiglietti, Global Water Mass: GRACE Satellite Monthly Data 2002–11, accessed September 28, 2013, www.visualizing.org/datasets/global-water-mass-grace-satellite-monthly-data-2002-11.
14. In this book I use the term *watershed* to refer to the area of land draining to a specific point, such as a lake or a point along a river, as described more fully in chapter 2.

Other commonly used equivalents include *catchment* and *river basin*.

15. Asher Price, "Farmers Battle State Environmental Agency in Brazos River Basin Dispute," *Austin American-Statesman*, December 26, 2012, accessed September 28, 2013, www.statesman.com/news/news/state-regional/farmers-battle-state-environmental-agency-in-brazo/nTf9w/.
16. World Economic Forum, *Global Risks 2012* (Geneva).
17. *Examples of the Impacts of Water Shortages*, Conservation Gateway, last updated April 29, 2013, www.conservationgateway.org/Files/Pages/examples-economic-impacts.aspx.
18. The Transboundary Freshwater Dispute Database, accessed September 28, 2013, www.transboundarywaters.orst.edu/database/DatabaseIntro.html.
19. The World Bank and State Environmental Protection Administration of China, *Cost of Pollution in China: Economic Estimates of Physical Damages* (Washington, DC: The World Bank, 2007).
20. Sandra Postel and Brian Richter, *Rivers for Life: Managing Water for People and Nature* (Washington, DC: Island Press, 2003).
21. World Economic Forum, *The Bubble Is Close to Bursting: A Forecast of the Main Economic and Geopolitical Water Issues Likely to Arise in the World during the Next Two Decades* (Geneva: 2009).

Chapter 2

1. International Water Management Institute, *Water for Food, Water for Life*, ed. David Molden (London: Earthscan, 2007).
2. It is important to acknowledge that insufficient access to water can also be caused by social inequities in the way that water supplies are distributed, lack of permission to access certain water sources, inabilities to pay for water, or pollution of water sources that renders them unusable. These biological, social, legal, and economic impediments are left to later chapters while I focus on physical scarcity here.
3. Water can also be imported from other watersheds or aquifers. These "interbasin transfers" of water are addressed in chapter 3.
4. U.S. Bureau of Reclamation, *Colorado River Basin Water Supply and Demand Study* (Denver: U.S. Bureau of Reclamation, 2012).
5. Until 1995, the U.S. Geological Survey reported both withdrawals and consumption, but because of budget constraints, they stopped reporting water consumption in recent years. Figure 7 of their 1995 report (Wayne B. Solley, Robert R. Pierce, and Howard A. Perlman, *Estimated Use of Water in the United States in 1995* (Washington, DC: U.S. Geological Survey, 1998) is one of the most useful water graphics I have ever seen, but it is now out of date.
6. I will use the terms *consumptively used*, *consumed*, and *depleted* interchangeably throughout the rest of the book.
7. Edwin D. Gutentag, Frederick J. Heimes, Noel C. Krothe, Richard R. Luckey, and John B. Weeks, *Geohydrology of the High Plains Aquifer in Parts of Colorado, Kansas, Nebraska, New Mexico, Oklahoma, South Dakota, Texas, and Wyoming* (Washington, DC: U.S. Geological Survey, 1984); and Virginia L. McGuire, *Changes in Water Level and Storage in the High Plains Aquifer, Predevelopment to 2009* (Denver, CO: U.S. Geological Survey, 2011).

Chapter 3

1. Edward Wong, "Plan for China's Water Crisis Spurs Concern," *New York Times*, June 1, 2011.

2. This ordering of relative costs is generalized and may differ in specific settings.
3. Erika Lee, "Saudi Arabia and Desalination," *Harvard International Review*, December 23, 2010, http://hir.harvard.edu/pressing-change/saudi-arabia-and-desalination-0.
4. "Solar-Powered Desalination Plant in Saudi to Be Completed by End of This Year," arabianindustry.com, August 26, 2013, http://arabianindustry.com/construction/news/2013/aug/26/solar-powered-desalination-plant-in-saudi-to-be-completed-by-end-of-this-year-4423276/.
5. Steve Maxwell, "Growing Awareness, Growing Risks," *2013 Water Market Review* (Boulder, CO: TechKNOWLEDGEy Strategic Group, 2013).
6. Committee on the Assessment of Water Reuse as an Approach to Meeting Future Water Supply Needs, Water Science and Technology Board, Division on Earth and Life Studies, and National Research Council, *Water Reuse: Potential for Expanding the Nation's Water Supply through Reuse of Municipal Wastewater* (Washington, DC: National Academies Press, 2012).
7. This is also commonly referred to as "before Christ," or BC.
8. The Water Footprint Network provides a catalog of water footprint estimates for many different consumer products on their website: www.waterfootprint.org.
9. Brian D. Richter, Sandra Postel, Carmen Revenga, Thayer Scudder, Bernhard Lehner, Allegra Churchill, and Margaret Chow, "Lost in Development's Shadow: The Downstream Human Consequences of Dams," *Water Alternatives* 3 (2010): 14–42.
10. Brian D. Richter, David Abell, Emily Bacha, Kate Brauman, Stavros Calos, Alex Cohn, Carlos Disla, Sarah Friedlander O'Brien, David Hodges, Scott Kaiser, Maria Loughran, Cristina Mestre, Melisssa Reardon, and Emma Siegfried, "Tapped Out: Growing Cities in Search of the Next Oasis," *Water Policy* 15 (2013): 335–363.
11. U.S. Environmental Protection Agency, *Water Smart Landscapes* (Washington, DC: 2013).
12. See note 10.

Chapter 4

1. Thomas L. Friedman, "Without Water, Revolution," *New York Times*, May 18, 2013.
2. For further discussion of different governmental approaches to water ownership and allocation, see Salman M. A. Salman and Daniel D. Bradlow, *Regulatory Frameworks for Water Resources Management: A Comparative Study* (Washington, DC: The World Bank, 2006).
3. A variety of terms are used around the world for describing a right to use water. The most commonly used terms are *entitlements*, *licenses*, *concessions*, *permits*, and *allocations*. For consistency, I will use *entitlements* in this book.
4. World Economic Forum, *Water Security: The Water-Food-Energy-Climate Nexus* (Washington, DC: Island Press, 2011).
5. Leonard F. Konikow, *Groundwater Depletion in the United States (1900–2008)* (Washington, DC: U.S. Geological Survey, 2013), available only online at http://pubs.usgs.gov/sir/2013/5079/.
6. The winter 2005 issue of *Cultural Survival Quarterly* includes a number of papers discussing these basic human rights and their implementation.
7. For further discussion of community-based water law, see *Community-based Water Law and Water Resource Management Reform in Developing Countries*, eds. Barbara van Koppen, Mark Giordano, and John Butterworth (Oxfordshire, UK: CAB International, 2007).
8. Eleanor P. Dictaan-Bang-oa, "Traditional Water Management Practices of the

Kankanaey," *Cultural Survival Quarterly* 29.4 (2005), accessed December 5, 2013, www.culturalsurvival.org/publications/cultural-survival-quarterly/philippines/traditional-water-management-practices-kankanaey.

9. U.S. Environmental Protection Agency, "EPA Survey Shows $384 Billion Needed for Drinking Water Infrastructure by 2030," June 4, 2013, accessed September 28, 2013, http://yosemite.epa.gov/opa/admpress.nsf/0/F72C2FDC7D61F92085257B800057655F.
10. U.S. Geological Survey, "Trends in the Size of the USGS Streamgaging Network," http://water.usgs.gov/nsip/streamgaging_note.html.
11. Global Renewable Fuels Alliance, "62 Countries Now Have Biofuels Friendly Policies—GRFA Online Interactive Map Shows Growth of Global Biofuels Policies," August 22, 2013, http://globalrfa.org/news-media/62-countries-now-have-biofuels-friendly-policies-grfa-online-interactive-map-shows-growth-of-global-biofuels-policies.
12. Robert F. Service, "Another Biofuels Drawback: The Demand for Irrigation," *Science* 23 (2009): 516–517.
13. Dan McCool, *River Republic* (New York: Columbia University Press, 2012).

Chapter 5

1. Peter H. Gleick, "Basic Water Requirements for Human Activities: Meeting Basic Needs," *Water International* 21 (1996): 83–92.
2. Brian D. Richter, Mary Davis, Colin Apse, and Christopher Konrad, "A Presumptive Standard for Environmental Flow Protection," *River Research and Applications* 28 (2012): 1312–1321.
3. Arjen Y. Hoekstra, Mesfin M. Mekonnen, Ashok K. Chapagain, Ruth E. Mathews, and B. D. Richter, "Global Monthly Water Scarcity: Blue Water Footprints versus Blue Water Availability," *PLoS ONE* 7 (2012), www.plosone.org/article/info%3Adoi%2F10.1371%2Fjournal.pone.0032688.
4. See note 7 of chapter 4.
5. "Experts: Water Diversion from Bohai to Xinjiang Unfeasible," People's Daily Online, November 17, 2010, http://english.peopledaily.com.cn/90001/90776/90882/7201757.html.
6. Stanley Crawford, *Mayordomo: Chronicle of an Acequia in Northern New Mexico* (Albuquerque: University of New Mexico Press, 1988).
7. For a fuller analysis of water policy changes and the conditions that led to them, see Sander Meijerink and Dave Huitema, "Policy Entrepreneurs and Change Strategies: Lessons from Sixteen Case Studies of Water Transitions around the Globe," *Ecology and Society* 15 (2010), www.ecologyandsociety.org/vol15/iss2/art21/.

Chapter 6

1. William Finnegan, "Leasing the Rain," *The New Yorker*, April 8, 2002.
2. See for instance the movie *FLOW: For Love of Water* or the book *Blue Gold: The Fight to Stop the Corporate Theft of the World's Water* by Maude Barlow and Tony Clarke (New York: The New Press, 2005). The plot of the James Bond 007 movie *Quantum of Silence* was also inspired by the Cochabamba incident.
3. This consortium was called Aguas del Tunari, composed of engineering and construction firms including Bechtel Corporation (USA), International Water Limited (England), Abengoa (Spain), Edison (Italy), and two companies from Bolivia, ICE Ingenieros and the cement maker SOBOCE.
4. See note 1.
5. "Bechtel Perspective on the Aguas del Tunari Water Concession in Cochabamba,

Bolivia," Bechtel, March 16, 2005, www.bechtel.com/2005-03-16_38.html.

6. Juan Forero, "Who Will Bring Water to Bolivia's Poor?," *New York Times*, December 15, 2005.
7. 2011 South Central Texas Regional Water Plan, www.regionltexas.org/2011_rwp2.php.
8. Katherine A. Roach, "Texas Water Wars: How Politics and Scientific Uncertainty Influence Environmental Flow Decision-Making in the Lone Star State," *Biodiversity Conservation* 22 (2013): 545–565.
9. This is based on expert testimony and evidence submitted as part of the federal lawsuit, and ensuing personal communications with Joe Trungale and Jim Blackburn.
10. For those unfamiliar with this celebration, it takes place on February 14 each year.
11. Andrew Cawthorne and Jack Kimball, "Valentine's Day Roses Bloom amid Kenyan Poverty," Reuters Planet Ark, February 13, 2006, www.planetark.com/dailynewsstory.cfm/newsid/35048/story.htm.
12. Richard Fox, Imarisha Naivasha: A Public Sector–Private Sector–People Participatory Approach to Resource Management, presentation by Richard Fox of Finlay's Horticulture Ltd, Kenya.
13. Lewis Smith, "Why Fruit Shopping Poisons the Hippo," *The Times*, March 28, 2006, www.thetimes.co.uk/tto/news/world/article1970065.ece.
14. Food & Water Watch and Council of Canadians, *Lake Naivasha: Withering under the Assault of International Flower Vendors* (Washington, DC: 2008).
15. Republic of Kenya, *Water Allocation Plan—Lake Naivasha 2010–2012* (Nairobi: Water Resources Management Agency, 2010).
16. See note 15.

Chapter 7

1. Donna Green, Jack Billy, and Alo Tapim, "Indigenous Australians' Knowledge of Weather and Climate," *Climate Change* 100 (2010): 337–354.
2. Much of the historical information discussed here is from Daniel Connell, *Water Politics in the Murray-Darling Basin* (Annandale, NSW: Federation Press, 2007).
3. Murray-Darling Basin Authority, *Guide to the Proposed Basin Plan* (Canberra: 2010).
4. Much of the water and agricultural data in this chapter comes from a report prepared by the Commonwealth Scientific and Industrial Research Organisation of Australia, *Water Availability in the Murray-Darling Basin* (Canberra: 2008).
5. See note 3.
6. Murray-Darling Basin Ministerial Council, *An Audit of Water Use in the Murray-Darling Basin* (Canberra: 1995).
7. Murray-Darling Basin Authority, *The Cap Brochure*, available at www.mdba.gov.au/what-we-do/managing-rivers/the-cap.
8. New South Wales Office of Water, *Water Reform in the NSW Murray-Darling Basin: Summary of Regional Water Reform and Environmental Water Recovery in NSW 1996–2011* (Sydney: 2011).
9. The final Basin Plan published in 2012 by the Murray-Darling Basin Authority reported that as of 2009, 13.6 billion cubic meters of entitlements remained on the books. The volume of entitlements was likely higher at the onset of the Millennium Drought in the late 1990s, because the volume of entitlements was reduced by environmental water buybacks during the drought.
10. Murray-Darling Basin Authority, *The Living Murray Story: One of Australia's Largest River Restoration Projects* (Canberra: 2011).
11. The rainfall summary presented here is based on N. J. Potter, F. H. S. Chiew, A. J. Frost,

R. Srikanthan, T. A. McMahon, M. C. Peel, and J. M. Austin, *Characterisation of Recent Rainfall and Runoff in the Murray-Darling Basin: A Report to the Australian Government from the CSIRO Murray-Darling Basin Sustainable Yields Project* (Canberra: CSIRO, 2008).

12. Peter Davies, John Harris, Terry Hillman, and Keith Walker, *Sustainable Rivers Audit: A Report on the Ecological Health of Rivers in the Murray-Darling Basin, 2004–2007* (Canberra: Murray-Darling Basin Commission, 2008).
13. Much of this economic summary is based on a paper by Mac Kirby, Jeff Connor, Rosalind Bark, Ejaz Qureshi, and Scott Keyworth, *The Economic Impact of Water Reductions during the Millennium Drought in the Murray-Darling Basin*, presented at the Australian Agricultural & Resource Economics Society's 56th annual conference, Fremantle, Western Australia, February 7–10, 2012), http://ageconsearch.umn.edu/bitstream/124490/2/2012AC%20Kirby%20CP%20of%20Poster2.pdf.
14. See note 3.
15. In October 2012, Prime Minister Julia Gillard announced that another $1.77 billion would be appropriated for the purpose of acquiring another 450 million cubic meters of water (365,000 acre-feet) for environmental purposes. This will bring the total volume of buybacks to 3.2 billion cubic meters (2.6 million acre-feet).
16. See note 12.

Chapter 8

1. World Water Council, *World Water Vision: Making Water Everybody's Business*, www.worldwatercouncil.org/index.php?id=961.
2. For more information about the Coca-Cola Company's community water program, see *The Water Stewardship and Replenish Report 2012*, http://assets.coca-colacompany.com/8d/d8/8f1cc9e3464e8b152f97aa91857b/TCCC_WSRR_2012_FINAL.pdf.
3. More information about the Stockholm Industry Water Award and the accomplishments of Netafim can be found online at www.siwi.org/prizes/stockholm-industry-water-award/winners/2013-2/.
4. More information about the United Nations CEO Water Mandate is available online at http://ceowatermandate.org/.
5. United Nations CEO Water Mandate, *Guide to Water-Related Collective Action* (New York: United Nations, 2012).

About the Author

Brian Richter has been a global leader in river science and conservation for more than 25 years. He is the director of Global Freshwater Strategies for The Nature Conservancy, an international conservation organization, where he promotes sustainable water use and management with governments, corporations, and local communities. Brian has consulted on more than 120 water projects worldwide. He serves as a water advisor to some of the world's largest corporations, investment banks, and the United Nations and has testified before the U.S. Congress on multiple occasions. He also teaches a course on Water Sustainability at the University of Virginia. Brian has developed numerous scientific tools and methods to support river protection and restoration efforts, including the *Indicators of Hydrologic Alteration* software that is being used by water managers and scientists worldwide. Brian was featured in a BBC *Horizon* documentary with David Attenborough, "How Many People Can Live on Planet Earth?" He has published many scientific papers on the importance of ecologically sustainable water management in international science journals, and he coauthored an earlier book with Sandra Postel entitled *Rivers for Life: Managing Water for People and Nature* (Island Press, 2003).

Index

Page numbers followed by "f" and "t" indicate figures and tables.

Island Press | Board of Directors